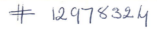

Microstructural Control in Aluminum Alloys:

Deformation, Recovery and Recrystallization

Front Cover
Transmission electron micrograph of Al-0.5% Fe-0.5% Co alloy
showing second phase particles pinning subgrains after
annealing.

ML

This book is to be returned on or before
the last date stamped below.

DEC 1990

Microstructural Control in Aluminum Alloys:

Deformation, Recovery and Recrystallization

Proceedings of a symposium sponsored by the Non-Ferrous Metals Committee of The Metallurgical Society, held at the Annual Meeting of TMS in New York, New York, February 27, 1985.

Edited by:

E. Henry Chia
Southwire Company
Carrollton, Georgia

H. J. McQueen
Concordia University
Montreal, Canada

A Publication of The Metallurgical Society, Inc.

Library of Congress Cataloging-in-Publication Data
Main entry under title:

Microstructural control in aluminum alloys.

Includes index.
1. Aluminum alloys--Congresses. 2. Microstructure--Congresses. I. Chia, E. Henry.
II. McQueen, H.J. III. TMS-AIME Nonferrous Metals Committee. IV. Metallurgical
Society of AIME. Meeting (1985 : New York, N.Y.)
TN693.A5M53 1986   669'.96722   85-29842
ISBN 0-87339-013-X

A Publication of The Metallurgical Society, Inc.
420 Commonwealth Drive
Warrendale, Pennsylvania 15086
(412) 776-9000

Printed in the United States of America.
Library of Congress Catalogue Number   85-29842
ISBN NUMBER   0-87339-013-X

© 1986

D
669.9672'2
MIC

PREFACE

One of the most important facets in the physical metallurgy of aluminum alloys is the study and control of microstructural features which are present during the different stages of fabrication. Direct applications of improved thermomechanical treatments during fabrication have allowed metallurgists to greatly benefit from the proper manipulation of the second phases and subgrain structure in order to obtain improved mechanical and electrical properties. This can be illustrated by the fact that impurities present in aluminum which were considered undesirable years ago are germane to the alloy's properties when present in the proper size and distribution.

This symposium has been able to cover the latest advances in the microstructural control during deformation, recovery and recrystallization processes, with emphasis on hot rolling and extrusion operations.

We are very thankful to the contributing authors for their efforts in the presentations and their valuable contributions to this subject.

E. Henry Chia
Southwire Company
Carrollton, GA

H. J. McQueen
Concordia University
Montreal Canada

TABLE OF CONTENTS

THE MICROSTRUCTURAL STRENGTHENING MECHANISMS

IN DILUTE Al-Fe CONDUCTOR ALLOYS

H.J. McQueen*, H. Chia** and E.A. Starke***

*   Prof., Mechanical Eng., Concordia University, Montreal, Canada H3G 1M8
**   Research Metallurgist, Southwire Company, Carrollton, Georgia, 30119
***   Dean of Engineering and Applied Science, University of Virginia,
       Charlotteville, VA, 22903

Abstract

Al alloys with 0.5-0.9% Fe content have largely supplanted 1350 EC alloy for building electrical circuits because the latter frequently suffered from gradual loosening at terminals, which gave rise to overheating. This problem has been completely overcome in the new conductor alloys without sacrifice of conductivity. The strengthening mechanisms, which give rise to the required combination of strength, bendability and softening resistance, are a retained dynamically-recovered substructure reinforced and stabilized by dispersed 0.2 μm, eutectic rods.

Introduction

Aluminum 1350 Electrical Conductor alloy with a conductivity 61% of International Annealed Cu Standard has a mass resistivity (0.0764 Ω for a 1 gm conductor 1 m long) which is only one half that of Cu (0.15328 Ω g/m^2). The primary application is electrical distribution in buildings with potential in automobiles, aerospace, telephone lines and magnet winding. To economically realize the weight advantage, aluminum wire must be capable of attaching securely to standard fixtures without special handling techniques (1,2). However, EC wire on binding screw terminals tightened to a standard torque can become loose. When the connection heats slightly due to a minor overload, the wire expands more than the Cu-alloy fixture and creeps to relax the added stress. On cooling it contracts to a smaller dimension, thus reducing the area of contact and allowing oxide to form at the interface. On a subsequent large current flow, the overheating increases so that additional plastic flow and loosening occur to further diminish the integrity of the joint (1). EC wire, annealed for adequate bendability, has a substructure softening at 200°C and consequently fails due to repetition of such cycles.

The new alloys (8000 series) of 0.5-0.9% Fe, such as Triple E, have greatly improved microstructural stability and creep resistance (Figure 1) and are not subject to such junction failure (3). At 180°C, the strength of annealed Triple E declines from 125 to 116 MPa in 500 hrs and to 100 in 2000 hrs, whereas EC-Al falls rapidly to 104 and 82 MPa, respectively. When annealed to the same ductility or bendability, the high Fe alloys are about twice as strong (Figure 2). This capability has been confirmed by field trials of several years in the United States, Europe and South Africa after these alloys were introduced in 1968 (1,4). More advanced alloys, which not only give high integrity to terminations but are suitable for magnet wire after

1

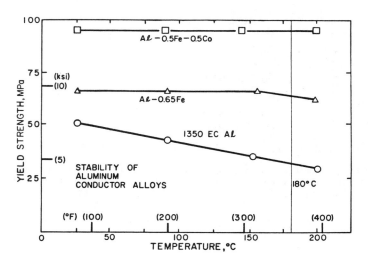

Figure 1   Effect of 100 hour high temperature stability test on the yield strengths of conductor alloys: 1350 EC Al, Triple E (Al-0.65Fe) and Super T (Al-0.5Fe-0.5Co) (3).

standard hot enameling, have been developed by additions of a third element to improve the distribution and morphology of the stabilizing particles; examples are 0.5% Fe with 0.5% Co (Super T) and 0.5% Fe with 0.2-0.4% Si (Almhoflex). These improvements have been attained without sacrificing conductivity since the eutectic phases, $FeAl_3$ (transformed from $FeAl_6$), $(Fe,Co)Al_3$, $(Co,Fe)_2Al_9$ and Fe_2SiAl_8 increase the resistivity very little (only 0.58 μohm-cm/wt% Fe compared with 2.56 μohm-cm/wt% Fe for Fe in solution) (5).   The development and patenting of such high Fe additions were remarkable since industry generally considered Fe an impurity to be limited to less than 0.4% (4,5). These alloys, with operating tensile strengths of 110-130 MPa, are not intended to compete against high strength wire for overhead lines such as 6201 alloy (0.7 Si, 0.8 Mg) in which 305-330 MPa are attained with a sacrifice in conductivity to 54% IACS (2).

The objective of this paper is to examine the microstructural mechanisms which give the dilute Al-Fe alloys, such as Triple E, Super T and Almhoflex, their characteristic mechanical properties.   These mechanisms can most easily be understood from two viewpoints, strengthening and stabilizing; the distinction between these are valid since some strengthening mechanisms are unstable at medium temperatures (~ 200°C, 0.5 Tm °C).   Moreover, the strengthening and stabilizing mechanisms are actualized in that order by the processing, i.e. the first group during bar casting, rod rolling and wire drawing and the second during final annealing.   Each mechanism is discussed as if it were additive, although strictly the combined effect may be less than the sum of the individual contributions.   Supporting evidence from other alloy systems is also cited.

Processing and Microstructure

In continuous casting, as typically practiced on the Southwire Continuous Rod System (SCRS), a bar of about 50 cm² is produced at 16 m/min on a 2.5 m diameter copper wheel.   The rapid solidification results in a 20 μm dendrite arm spacing and eutectic rod spacing of about 0.2 μm (Figure 3) with a supersaturation of about 0.1% Fe.   As explained later these very fine parti-

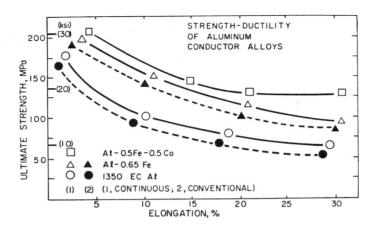

Figure 2   The relationship between ultimate tensile strength and ductility of conductor alloys annealed after drawing (elongation on 254 mm (10 in) gage length) (3). In continuous processing, the rod proceeds directly to wire drawing without a recrystallizing anneal, whereas in conventional, it is given such a treatment.

cles play an important role in stabilizing the substructure while being incapable of nucleating recrystallization (6-8).

The bar, direct from the casting wheel, is reduced 98.6% to 0.7 cm² in a 13 stand rolling mill. The cast structure is broken down as the grains are elongated and reduced in thickness to about 2.5 μm (Figure 4). Within them, a substructure is formed which becomes progressively refined as the temperature declines from 485 to 180°C. The eutectic colonies, extended and dispersed longitudinally, are brought closer together transversely. The particles are rotated, bent and fractured into shorter segments. The rod is cooled before coiling to avoid nonuniform subgrain coalescence (6-8). In continuous practice, the rod goes directly to wire drawing without intermediate annealing. As a result the hot-worked substructure, instead of statically recrystallized grains, is carried into the drawing process. Passage through the twelve dies reduces the rod another 92.2% to 0.052 cm². The cold worked wire is annealed at 250°C to restore its ductility to 15-20%.

Strengthening by Strain Hardening in Hot Working

Strain hardening in hot working decreases in rate as temperature T rises but increases as strain rate $\dot{\varepsilon}$ rises. The lower rate compared to cold working is the result of dynamic recovery, i.e. reduced accumulation of dislocations because of annihilations and combinations. Dislocations arrange themselves into subboundaries with neater, less ragged arrays and larger link lengths as T rises and $\dot{\varepsilon}$ falls; the spacing of the arrays, the subgrain size, also becomes larger (Table I). The hot strength σ_s is inversely proportional to subgrain diameter d_{sg} (9-15). Substructure strengthening is also produced by creep; the lower strain rate than that during hot-working gives rise to larger subgrains, up to 50 μm (16,17).

The presence of subgrains in hot worked aluminum has been known for some time but without quantitative determination of the dimensions or the effects on properties (18-23). As temperature rises from 200-450°C, the cold yield strength of the hot worked product decreases from 80% to 2% of the strengthening produced by 97.5% cold rolling (24). As has been observed in many hot

3

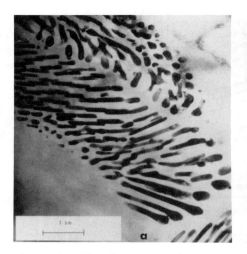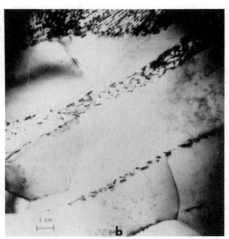

Figure 3   TEM micrograph of Aℓ-0.5Fe-0.5Co showing $(FeCo)_2Al_9$ eutectic rods in Aℓ matrix: a) as-cast and b) after 1 rolling pass to 37.3% red (7).

TABLE I   STRENGTH OF HOT WORKED ALUMINUM ALLOYS

ALLOY	T °C	$\dot{\varepsilon}$ STRAIN RATE s^{-1}	ε STRAIN	d_{sg} μm	σ_s HOT STRENGTH MPa	σ_y YIELD STRENGTH MPa
1100 (a) Al	200° 200° 400° 400° 475°	220 22 12 1.3 0.05	0.7 0.7 0.7 0.7 0.7	1.13 1.2 2.0 2.4 6.4	110 105 57 43 23	105 100 63 55 50
Al (b) 4.5 Mg 0.8 Mn	300 400 500	1.0 1.0 1.0	4 4 4	1.1 2.0 3.5	185 130 75	
0.65 Fe (c) Triple E	485- 180	1.6 27	4.33 13 passes	5.9 1.0	rod	110
0.5 Fe (d) 0.5 Co Super T	485- 180	1.6 27	4.33 13 passes	5.5 0.85	rod	120 d
0.6 Fe (e) 0.3 Si Almhoflex	450 350		3.58	0.6		
EC (d) 1350	485- 180	1.6 27	4.33 13 passes	1.6	rod	96

(a) - McQueen and Hockett 1970
(b) - McQueen et al 1984
(c) - Chia, Spooner and Starke 1977
(d) - Chia and Starke 1977
(e) - Fiorini 1970

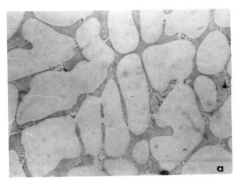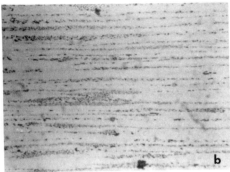

Figure 4 Optical micrographs of bar and rod showing eutectic particles in Aℓ-0.5Fe-0.5Co (Super T) a) as-cast and b) rolled rod, longitudinal x 450 (7).

TABLE II MECHANICAL AND ELECTRICAL PROPERIES OF Aℓ-Fe CONDUCTOR ALLOYS

ALLOY	d_{sg} μm	CONDITION	YIELD STRENGTH MPA	ULTIMATE STRENGTH MPA	EL %	IACS %	THERM STAB+ %
0.65 Fe(a) Triple E	1.0	rod	110	125	20	60.4	
	0.6	drawn wire	175	212	3	60.1	
	1.4	290°C, 3hrs annealed (b)	70	108	26	61.2	92
	1.75	2000 h 180°C (d)	90	110	27	62.2	
0.5 Fe (b) 0.5 Co Super T	0.85	rod	120	145	18	59.8	
	0.4	drawn wire	180	212	3	59.1	
	0.8	300°C, 1 hr	131				
	0.7	290°C, 3 hrs	120	140	18	61.4	98
		annealed (b)	102	132	18		
	1.0	2000 h 180°C (d)	99	120	25	62.2	
0.6 Fe (c) 0.3 Si Almhoflex	0.6	rod					
		drawn wire		233	4		
		annealed (c)		130	25		
EC (ab) 1350	1.6	rod	96	110	15	61.9	
	0.8	drawn wire	160	173	2	61.6	
	1.6	300°C, 1 hr	87				
	2.8	290°C, 3 hrs	58	88	26		71
		annealed (b)	105	124	4		
		annealed (c)		130	8		
	4.6	2000 h 180°C (d)	60	82	25	63.0	
UL Laboratories Requirement			76	103	10		

Rod rolled through 13 passes ε_R = 4.33 Wire drawn through 12 dies ε_W = 2.56

a - Chia Spooner and Starke 1978   d - Leneaus and McPheters 1972
b - Chia and Starke 1977   e - Schoerner 1975
c - Fiorini 1970 (ε_R = 3.58, ε_W = 3.33) + strength retained after 4 hrs, 250°C

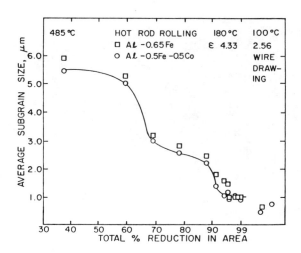

Figure 5   Subgrain size as a function of reduction in cross-section for
Al-0.65Fe (Triple E) and Al-0.5Fe-0.5Co (Super T) (6,7).

worked metals, the yield strength σ_y is inversely proportional to subgrain diameter d_{sg} (11,13,17,25-29):

$$\sigma_y = \sigma_A + k\, d_{sg}^{-1} \tag{1}$$

where σ_A is the yield strength of metal with recrystallized grains considerably
larger than the subgrains and k is the subgrain strengthening coefficient.
Furthermore, substructure provides effective creep resistance if it is finer
than the subgrains characteristic of the creep conditions and if it is stabilized by particles to prevent evolution towards the latter (11,12,16,17).
However, the beneficial effects can also be catastrophically lost through
recrystallization during service.

During the multistage rod rolling (total $\varepsilon = 4.33$) of an Al-0.65% Fe
alloy, from 485°C down to 180°C with final $\dot{\varepsilon} \cong 27$ s^{-1}, the subgrain sizes after
successive passes decrease from 5.9 to 1.0 μm (Figures 5 and 6)(6,7).
Because the temperature is lower and the strain rate higher in a given pass
than those in the previous one, substructure "inherited" from, i.e. carried
forward from, the latter is altered by addition of dislocations to the existing
walls to raise their density and by formation of new walls to subdivide the
subgrains reducing their size. Because it takes considerable strain to effect
this conversion, the subgrain size after each pass remains larger than would be
expected solely from the deformation at the temperature and strain rate of the
pass (11,13,28,30,31). Furthermore, there is no opportunity for static
recrystallization between passes and little for static recovery as the time
decreases from 3 s to 0.06 s (7,29-33). Initially the subgrains form between
the eutectic colonies and later between the rows of eutectic rods. However as
the rods are dispersed and broken up by the deformation, they interact more
intimately with the dislocations as explained below. Because FeAl$_3$ particles
of 0.5 to 0.075 μm increase the dislocation density, thus decreasing the cell
size and raising the wall density and misorientation, the yield strength of the
Triple E rod is 110 MPa somewhat greater than that of 1100, or 1350, aluminum
(Table I). A hypothetical decrease in cell size from 1.1 μm to 0.9 μm and an
associated increase in wall density from 9x10^4 lines per cell face to 11x10^4
lines per cell face results in an increase in dislocation density from 2x10^{13}
to 3.6x10^{13} m/m^3; almost a two fold increase. Since the strength varies as the

Figure 6   TEM micro-
graphs of subgrains in
Aℓ-0.65Fe (Triple E) rod
at:
(a) intermediate and
(b) final rolling reduc-
     tion;
(c) in Aℓ-0.5Fe-0.5Co
     (Super T) final
     reduction and
(d) in EC Aℓ final reduc-
     tion (6,7).

square root of the dislocation density, the strength increases by a factor of
1.3. The 0.5 Fe-0.5 Co alloy has a higher density of particles and a subgrain
size of 0.85 μm (Figure 5 and 6) with a yield stress of 140 MPa (6,7). The
electrical conductivity is reduced by the dislocation substructure but, because
of the dynamic recovery, not as much as for metal cold worked to the same
strain. The conductivity of 0.65 Fe as rolled rod is 60.4% IACS whereas that
of 0.5 Fe-0.5 Co is 59.8% ACS (Table II).

Strain Hardening from Cold Drawing

Although in the past it was common practice to fully anneal the rod before wire drawing, continuous processing is now usually practiced, that is the as-worked rod with retained substructure goes directly to the cold forming. Drawing of the rod through 12 dies with 20% reduction per die brings the total true strain to 6.89 (6). Such additional strain of 2.56, about 60% of the hot rolling, requires only 60% of the dislocation motion, but brings about much more dislocation storage and a higher rate of strain hardening because of comparatively little dynamic recovery at ambient temperature (17,34). The Triple E wire strength is increased to 175 MPa up 60% from 110 MPa of the rod as a result of the cells decreasing to 0.6 μm (Figures 5 and 7) (6). In Super T, strength is somewhat greater and the cell size smaller (Table II). The greater strength of continuously treated wire compared to conventionally processed (Figure 2) results partly from finer cell sizes (3,6,7,35,36), e.g. after drawing and annealing they are 0.5 and 0.7 μm for continuous and 0.8 and 1.5 μm for annealed. The strain hardening in wire drawing of EC Al and some dilute alloys was linear when there was a retained hot worked substructure (35-38), but became zero or negative when the rod had recrystallized grains or coarse particle distributions (38,39). The softening is likely the result of dynamic recovery, not recrystallization (40,41), since no new grains were observed (9-12).

The cold worked cell structure (Figure 7) is built on the existing hot worked structure without tearing it apart. Dislocations become entangled in the existing sub-boundaries, thus making them more ragged and reducing the links of the wall networks, and also form new walls partitioning the subgrains and decreasing the cell size to 0.5-0.8 μm (somewhat similar to the substructure conversion required by changes of T or $\dot{\varepsilon}$ described above). This behavior is similar to that found upon reloading of a cold worked specimen after cell growth in static recovery; the flow curve is lower than the initial cold work curve of recrystallized material (at the same total strain) because dislocations are accumulating on the recovered substructure in a different way from on the purely cold worked (42). The hot worked structure in Aℓ-0.65 Fe is much more stable and less disturbed by the cold working than EC wire or commercial aluminum because of the stabilizing effect of the 0.2 μm FeAℓ₃ particles. The strengths of continuously processed Al and dilute alloys after wire drawing follows Equation 1 (35,36,38) and k is higher than that of recrystallized starting materials (35,36).

The cold working decreases the conductivity of the wire relative to the hot rolled rod because of the greatly increased dislocation density. The electrical conductivity of Triple E wire in the as drawn condition is 60.1% IACS while that of Super T is 59.1% IACS (Table II).

Particle Hardening

Primary particles of FeAℓ₃ greater than 0.6 μm diameter, are not shearable, i.e. dislocations cannot pass through them. As they do not change shape, the surrounding matrix flows around them undergoing additional, complex deformation. This results in creation of cells much smaller than the average size with dense, high misorientation walls (43-45). The needle-like, eutectic particles of FeAℓ₃, (Fe,Co)₂Aℓ₉, or Fe₂SiAℓ₃, of about 0.2 μm diameter are also not shearable by individual dislocations (6,7,46,47). The metal flows relatively easily around these thin rods, so that they give rise to merely additional dislocations (Figures 5, 6 and 7) (3,6,7,43-45). However, dislocations accumulating along their length, exert bending stresses that fracture them into segments only a little longer than their diameter.

On a larger scale, the extreme reductions and elongation of processing extends each eutectic colony into a long fiber of fractured particles and re-

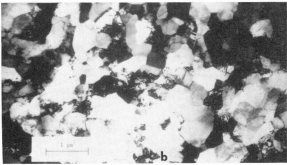

Figure 7   TEM, cellular
substructure in drawn wire
(a) Aℓ-0.65Fe (Triple E),
    and
(b) 1350 EC.

duces the separation between such fibers across the section making their distribution more uniform (Figure 4). Such high uniformity has been attained for
eutectic FeAℓ₃ particles in a chilled casting by extrusion to about 6.4 (48).
and for (Mn,Fe)Aℓ₆ rods in Aℓ-4.5Mg-0.8Mn by extrusion to 4.38 (49), or by
torsion to 5.5 whereby the majority of the rods have an aspect ratio between
2.5 and 1 (50). The above particles, particularly the fine ones at high volume
fraction increase the dislocation density and block dynamic recovery processes
leading to (i) finer cell sizes with higher density walls on the average and,
(ii) increased strain hardening and final strength compared to pure aluminum
(Figure 2). Triple E has smaller cells and denser walls than EC because of its
higher concentration of particles in the range 0.075-0.5 μm (Figures 7 and 8).
In Aℓ-Aℓ₂O₃ alloys, the cell size becomes finer, 0.7 to 0.6 μm, as the volume
fraction of Aℓ₂O₃ increases (51). In Aℓ-5Mg-0.8Mn, half of the particles are in
subboundaries and the remainder pin dislocations within the cells (50), with
the result that the subgrain diameters vary inversely with the temperature
compensated strain rate as do those of Aℓ, but are only half as large (49,50).
As the particle density increases in wires drawn from dilute alloys (Fe-Si,
Fe-Co, Ni, Fe-Cu, Fe-Mg), the cells are finer at a fixed strain and the alloys
stronger at the same cell size so that in Equation 1 the value of k is higher
(35-38). However, coarse poorly-distributed particles (due to slow solidification or inadequate hot rolling) lead to work softening at high strains (38-39).

     Precipitation of very small FeAℓ₃ particles($\cong$ 0.01 μm) takes place as the
solid solubility of Fe decreases from 0.052% at 655°C (probably much higher
because of rapid solidification) to only 0.005% at 500°C. In Aℓ-0.4% Fe particles form coherently either during slow cooling or on aging after quenching,
but gradually become noncoherent to give maximum hardness in 50 hrs at 350°C or
15 hrs at 390°. Precipitation is greatly accelerated by addition of 0.05% Si
or by plastic deformation after solution treatment (52-55). The age hardening
due to FeAℓ₃ is much less than in Aℓ-Cu alloys where the solubility is 5.65%

9

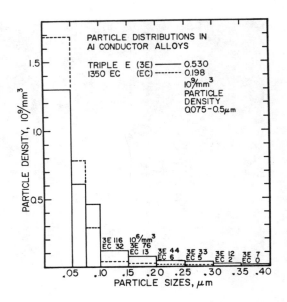

Figure 8   Particle distributions in Triple E and 1350 EC. The 0.05-0.10 μm category was split mathematically assuming two distinct populations of very fine and medium size particles.

because the volume fraction of $FeAl_3$ is about 1/100 that of $CuAl_2$ (52-53). Thus in the hot rolling of Al-0.65 Fe rod, precipitation takes place mainly on the dislocations because of the gradual cooling. Such small particles lock individual dislocations in the sub-boundaires decreasing dynamic recovery and increasing dislocation density and strain hardening (56,57). However, some precipitation has been observed inside the cells (6,54). In Almhoflex the excess Si precipitates as 0.01-0.02 μm particles of α (Fe_2SiAl_8) (46,47).

Dispersions of insoluble particles are very effective in strengthening at high temperatures both in creep where it is desired and in hot working where it makes forming difficult (11,12,16,17,23,58,59). At elevated temperatures the dislocations can bypass the particles by cross slip or climb; nevertheless, this slows down the recovery processes and raises the stress required for a given strain rate or lowers the creep rate produced by a given load. The above particles are incoherent and sufficiently large to have little significant effect on the conductivity other than through increasing the dislocation density. However, they cause a major indirect reduction by lowering the amount of Fe or Co in solid solution (Table II).

Solid Solution Hardening and Stabilization

The potential solution hardening by Fe is high because of the large atomic size difference of 5.9%; however, the actual hardening depends on the amount dissolved (5). In a 0.5% Fe alloy, the strength after quenching from 640° to retain 0.05% Fe in solution and severely deforming is about 190 MPa compared to 170 MPa for furnace cooling for precipitation. A 0.05% Fe alloy similarly treated has strengths of 170 and 110 MPa respectively. These strengths increase slightly upon low temperature aging as solute segregates to the dislocations but decline to about 40 MPa at 310°C as precipitation and recrystallization take place (60). In the as-cast Triple E alloy, there is considerable solute hardening since the concentration may be as high as 0.1% Fe (6). However, precipitation during hot rolling returns the concentration to equilibrium which at the finishing temperature is at such a low level that it provides very little hardening, even though the Fe atoms form weak Cottrell atmospheres (60). The levels of Fe in solution and the hardening are likely to be no different in Aℓ-0.65 Fe and in EC (~0.2 Fe) similarly processed. Marked solute strengthening in 0.7 Fe alloys is achieved by addition of as little as 0.2 Mg, or Cu (35,36).

Solid solution of Fe at the saturation level is so low that it does not affect the ability of the dislocations to recover by effects such as change of stacking fault energy or dynamic strain aging. Furthermore, the little Fe present as solute does not appreciably slow down recrystallization through segregation to the grain boundaries (54). The final annealing raises the conductivity by completing the precipitation (Table II).

Strength and Stability from Grain Size and Shape

In the course of rolling and wire drawing, the grains become fibrous, lengthening by a factor of 100 in the direction of the wire axis and decreasing in diameter by a factor of 10 in the plane normal to it; the rows of eutectic particles are reduced in spacing to about 2 μm. The longitudinal strength is raised because the area for glide of mobile dislocations in slip planes diagonal to the axis is much restricted. The grains in the EC conductor wire are also drawn out; however, the rows of particles are mainly absent. In the dilute Aℓ-Fe alloys, the type of rolling passes and the large volume of eutectic lead to a [111] fiber texture which is a much weaker mechanically than normal Aℓ alloy rod (5). As a result, the wire is more isotropic than otherwise would be the case and hence better suited for making pressure connections. Grain size strengthening may diminish over time at high temperature due to growth of the grains; however in the drawn Al-Fe alloys, growth is almost eliminated by the eutectic particles strung along the boundaries of the elongated grains.

Resistance to Softening of a Hot Worked Substructure

Strengthening from cold working has very low stability at elevated temperatures because the high density, high energy substructure readily gives rise to recrystallization unless some additional factor blocks it and provides an opportunity for recovery to gradually lower the strain energy and improve the stability. On the other hand, since Aℓ is highly capable of recovery, limiting its degree is important in maintaining strength. In the first stage of recovery, tangles diminish in density and rearrange into neat sub-boundaries as redundant dislocations annihilate with retention of the substructure scale and much of the strength (11,42,61). The polygonization in this stage is initially speeded up by the internal stresses in cell walls and interiors (61,62). In the second stage, the strength declines severely as subgrains become non-uniformly larger through walls either disintegrating as their dislocations leave to incorporate into others, or migrating to amalgamate with others. Such irregular coalescence can create recrystallization nuclei (7,9-13,35,37,40). The substructure in a metal worked at a high temperature is initially dynamically recovered to more regular, larger subgrains than a cold worked structure after first stage annealing at the working temperature. Nevertheless, holding without cooling causes the substructure to statically recover with loss of up to 40% of the strain hardening. For low hot prestraining (~10%), the recovery takes place to this level without ever recrystallizing (11,13,30-33,63). In recovery following creep in Aℓ-Zn alloys, the dislocation-density decrease in subgrain interiors and walls and the average subgrain-size increase occurs rapidly at first, augmented by the internal stresses, but later becomes very slow (54). Static recovery at 200°C of Aℓ hot worked at various higher temperatures decreases as the working temperature increases, thus confirming the greater stability of more dynamically recovered substructures (29). Moreover in annealing at a fixed temperature, specimens hot worked at lower temperatures are less resistant to recrystallization (65).

11

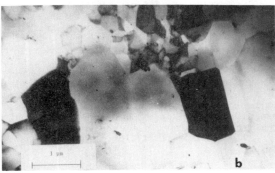

Figure 9   TEM subgrains
in annealed wire:
a) A ℓ-0.5Fe-0.5Co (Super
   T) and
b) EC 1350 (6,7).

In continuously processed alloy, where wire drawing has been imposed on a preexisting hot worked substructure, recovery starts with annihilation in the walls and cell interiors of redundant dislocations which were primarily added during cold deformation (Figure 9). The cold work tangles partitioning hot-worked subgrains disappear, partially by internal annihilations and partly by unravelling of dislocations attracted into the stable subboundaries (56). Just as the substructural alterations in cold drawing are less in A ℓ-0.65% Fe than in EC A ℓ, the underlying hot worked substructure of the former also decomposes less through recovery than EC due to the stability provided by the dispersed eutectic rods. Thus after process annealing, the remaining subboundaries in the continuously processed high Fe conductor wire are essentially those from the hot working. Moreover, since these were established initially at the entrance rather than the finishing temperature of the rod mill, they consist of neat low energy arrays which do not easily decompose in the coalescence process that gives rise to recrystallized nuclei. The stability of the preserved dynamically recovered substructure is much greater than that of a statically-recovered substructure in a wire drawn from recrystallized rod of a similar alloy (36,39,66). This contention is supported by the improved high temperature stability of A ℓ-Cu alloys hardened by repeated cycles of small cold deformations and recovery annealing compared to simple cold working (67). The drop in dislocation density during annealing raises the conductivity (Table II)

Fine Dispersion Stabilization

The fractured FeA ℓ$_3$ eutectic rods of about 0.2 μm diameter fairly uniformly distributed play an important role in stabilizing the substructure (Figure 1). The 10^9 mm^{-3} density of 0.075-0.5 μm particles in Triple E is the same as the number of 1 μm cells per mm^3, so that there is about one particle per cell (Figure 8); whereas for EC there is only one for every two cells.

12

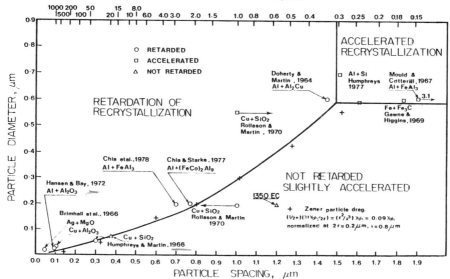

Figure 10 Summary of the effects of particles on recrystallization (6,7,43, 44,57,73-79). Only the particles sufficiently closely spaced, to the left side of the line (80-83), are able to effect retardation.

Lying mainly on sub-boundaries or at junctions thereof, the particles anchor the sub-boundaries sufficiently that the mechanisms of migration and unravelling for coalescence and recrystallization are much slowed down as has been observed for other dispersoids such as $MnAl_6$, $ZrAl_3$, and Al_2O_3 (56,68-75). In Al-5Mg-0.8Mn, $MnAl_6$ eutectic rods dispersed to subgrain corners, stabilize the substructure during hot extrusion thus limiting dynamic and static recrystallization (49). In Al-10% Fe atomized-powder extrusion-compacted alloy, the dendritic $FeAl_3$, uniformly distributed in particles of 0.3-0.03 μm, stabilizes the hot worked substructure and impedes recrystallization for up to 1000 hrs at 320°C (72). In combination with recovery annealing, θ' particles in Al-Cu alloys stabilize the substructure up to 400°C (67). Dilute dispersion alloys show good stability to work softening but coarse dispersions (0.7 Fe, 2-6 Ni) do not (38,39,41). Rapid, non-uniform subgrain growth was observed in EC Al during annealing (7,35,37,40).

The data for particle influence on discontinuous recrystallization, that is nucleation and growth by migration of a high angle boundary, is summarized in Figure 10 (6,7,43,44,57,73-79). To retard recrystallization, the necessary particle spacing decreases as the particle size decreases. The 0.65 Fe and the 0.5 Fe-0.5 Co alloy fall just within the stabilization region indicating that there is no excess of particles which could diminish the ductility and the conductivity. The possibility of recrystallization enhancement is discussed in the next section. Distributions in the region below the diagonal accelerate recrystallization slightly because they increase the dislocation density and strain energy. An alternative to nucleation prevention by substructure stabilization, is the prevention of growth through the Zener drag by particles of radius r and volume fraction f on migrating boundaries (43,50,80-84):

$$Z_d = \alpha(3f\,\gamma_{gb}/2r) \qquad (2)$$

However since the particle spacing λ equals $r(4\pi/3f)^{1/3}$

$$Z_d = \alpha\,2\pi\,\gamma_{gb}(r^2/\lambda^3) \qquad (3)$$

13

When the constants were evaluated for criticality at $\lambda = 0.8$ μm and $2r = 0.2$, μm, the limiting condition can be calculated for other particle sizes (32). The close agreement between the calculated points and the experimental line in Figure 10 indicates that the limit of retardation may be determined by the Zener inhibition of migration rather than by stabilization of the substructure.

The fine particles of $FeAl_3$, which precipitated on dislocations during working and cooling, cause local stabilization of the subboundary in which they are lodged, but are not in sufficient density to be generally effective as confirmed by their inability to cause stabilization of 1350 EC wire. This conclusion is not contradicted by certain observations of inhibition of discontinuous recrystallization through solution, quenching, cold working and aging because there the precipitated particles, such as $MnAl_6$, $FeAl_3$, $CuAl_2$ or $ZrAl_3$, are larger (54,80,85-87). In most of these cases, the high density of dislocations in the pinned walls leads to continuous recrystallization after extremely long annealing times (50,80,87).

Bimodal Stabilization

Large particles 0.6 μm in diameter have a destabilizing influence in so far as they serve as centers of nucleation because they have created around themselves local regions of fine subgrains with very high misorientations. Such large particles 0.6-2.5 μm widely spaced have been observed to accelerate recrystallization in Al-Fe alloys, as well as in several other alloy systems (44,45,49,76-78). The particles of less than 0.6 μm, such as the eutectic particles of 0.2 μm, do not act as centers of nucleation. In Al-Mn alloys, if Mn precipitates as large particles before deformation, the recrystallization temperature is lowered as the concentration of Mn increases; however if fine Mn particles form on dislocations during annealing, they stabilize substructure and inhibit recrystallization (86). In Al-0.39% Fe alloys, sand cast material has a lower recrystallization temperature than chill cast because of the coarser particles (88).

The stability of a bimodal particle distribution has been analyzed by Nes (31-83) who showed that suitable distribution of fine particles could prevent growth of nuclei formed at large particles. Such behaviour was observed in Al-Mn and Al-Mg-Si alloys with large particles of 1 μm from chill casting and in the former stabilization arose from 0.04 μm precipitates of $MnAl_6$ not from Mn in solution. This behaviour was confirmed for Al-Mn alloys with additional fine precipitates of either $ZrAl_3$ (71) or $MnAl_6$ (0.2-0.3 μm with spacing of 1-2 μm) (39). Similar inhibition was found in a commercial RR58 and special alloys where the large particles were Fe-Ni intermetallics and the fine, 0.2 μm spheres of MgCuSi (90). The above results explain the observations in Al-0.65% Fe that nuclei form at the large particles, but do not grow because of surrounding small particles. Regions containing 0.2 μm and smaller particles show no localized growth of subgrains as needed for recrystallization and of course continuing recovery gradually increases resistance to growth because the dislocation density continuously declines (56).

Conclusions

The Al-0.65 Fe and similar dilute Al-Fe alloys have greater strength combined with higher ductility than 1350 EC because they are strengthened by a hot worked substructure rendered more dense by a dispersion of fine eutectic rods resulting from rapid solidification. Such a substructure, retained from the hot rolling through continuous processing, and essentially unaltered by wire drawing and process annealing has inherently high stability which is enhanced by the particles of about 0.2 μm diameter.

REFERENCES

1. G.E. Leneaus and H.R. McPheters, Metals Materials, 6 (1972), pp. 401-403.

2. R. Iricibar, C. Pampillo, and H. Chia "Aluminum Transformation Technology and its Application, ASM, Metals Park, Ohio, (1978), pp. 241-303.

3. E.H. Chia, and E.A. Starke, Aluminium, 47 (1971), pp. 429-31.

4. R.J. Schoerner, Method of Fabricating Al Alloy Rod, U.S. Patent Re 28, 419, 13 May 1975.

5. K.R. van Horne, Aluminum, Amer. Soc. for Metals, Metals Park, Ohio (1967), Vol. 1, pp. 359-96.

6. E.H. Chia, S. Spooner, and E.A. Starke, Aluminium, 54 (1978), pp. 757-61.

7. E.H. Chia, and E.A. Starke, Met. Trans., 8A (1977), pp. 825-32.

8. E.H. Chia, Aluminum Transformation Technology and its Application, ASM, Metals Park, Ohio, (1978), pp. 305-33.

9. J.J. Jonas, C.M. Sellars and W.J. McG. Tegart, Met. Rev. 14 (1969), pp. 1-24.

10. C.M. Sellars, and W.J. McG. Tegart, Mem. Sci. Rev. Met., 63 (1966), pp. 731-46.

11. H.J. McQueen and J.J. Jonas, Plastic Deformation of Materials (Treatise on Mat. Sci. Tech., Vol. 6), Ed. R.J. Arsenault, Academic Press, New York (1975), pp. 393-493.

12. H.J. McQueen and J.J. Jonas, J. Appl. Metal Working, 3 (1984), pp. 233-41.

13. H.J. McQueen and J.J. Jonas, J. Appl. Metal Working, 3 (1984), pp, 410-20.

14. H.J. McQueen, W.A. Wong and J.J. Jonas, Can.J. Phys., 45 (1967), pp. 1225-34.

15. H.J. McQueen and J.E. Hockett, Met. Trans., 1 (1970), pp. 2997-3004.

16. O.D. Sherby, R.H. Klundt and A.K. Miller, Met. Trans., 8A (1977), pp. 843-50.

17. H.J. McQueen, Met. Trans., 8A (1977), pp. 807-24.

18. H.K. Hardy, Metallurgia, 30 (1944), pp. 240-44.

19. D. McLean and A.E.L. Tate, Rev. Mét., 48 (1951), pp. 765-75.

20. D. Hardwick and W.J.McG. Tegart, J. Inst. Met., 90 (1961), pp. 17-21.

21. H. Ormerod and W.J. McG. Tegart, J. Inst. Met., 92 (1963), pp. 297-99.

22. R. Leguet, D. Whitwham and J. Hérenguel, Mem. Sci. Rev. Met., 54 (1962), pp. 649-71.

23. H.J. McQueen, Metal Forum, 4 (1981), pp. 81-90.

24. W.A. Wong, H.J. McQueen and J.J. Jonas, J. Inst. Met., 93 (1967), pp. 129-37.

25. J.E. Hockett and H.J. McQueen, Proc. 2nd Int. Conf. Strength of Metals and Alloys, Am. Soc. Metals, Metals Park, Ohio (1970), pp. 991-95.

26. D.J. Abson and J.J. Jonas, Metal Sci., 4 (1970), pp. 24-28.

27. R.J. McElroy and Z.C. Szkopiak, Int. Met. Rev., 17 (1972), pp. 175-202.

28. H.J. McQueen, J. Met., 32 [2] (1980), pp. 17-26.

29. H.J. McQueen and W.G. Hutchison, Deformation of Polycrystals, RISO Natl Lab. Roskilde, Denmark (1931), pp. 335-42.

30. H.J. McQueen, Can. Met. Q., 21 (1982), pp. 445-60.

31. M.M. Farag, C.M. Sellars and W.J. McG. Tegart, Deformation Under Hot Working Conditions., Iron and Steel Inst., London (1968), pp. 60-67.

32. H.J. McQueen, E. Evangelista, J. Bowles and G. Crawford, Met. Sci., 18 (1984), pp. 395-402.

33. R.W. Evans and G.R. Dunstan, J. Int. Met., 99 (1971), pp. 4-14.

34. H.J. McQueen and W.J. McG. Tegart, Sci. Am., 232 [4] (1975), pp. 116-25.

35. D. Kalish and B.G. LeFevre, Met. Trans., 6A (1975), pp. 1319-24.

36. D. Kalish, G.G. LeFevre and S.K. Varma, Met. Trans., 8A (1977), pp. 204-06.

37. S.K. Varma, Res Mech., 3 (1983), pp. 175-83.

38. D.J. Lloyd and D. Kenny, Acta Met., 28 (1980), pp. 639-49.

39. S.K. Varma and B.G. LeFevre, Met. Tran., 11A (1980), pp. 935-42.

40. S.K. Varma, Scripta Met., 13 (1979), pp. 345-48.

41. S.K. Varma, Res Mech., 9 (1983), pp. 249-53.

42. T. Hasegawa and U.F. Kocks, Acta Met., 27 (1979), pp. 1705-16.

43. T.C. Rollason and J.W. Martin, J. Mat. Sci., 5 (1970), pp. 127-32.

44. F.J. Humphreys, Acta Met., 25 (1977), pp. 1323-44.

45. F.J. Humphreys, Deformation of Multi-Phase and Particle Containing Materials, ed. J. Bilde-Sorensen, RISO Natl. Lab, Roskilde, Denmark (1983), pp. 41-52.

46. P. Fiorini, Wire Industry, 46 (1979), pp. 265-68.

47. P. Fiorini, personal communication, Instituto Sperimentale dei Metalli Leggeri, Aluminio Italia, 23100 Novara Italy (1984).

48. J. Herenguel and P. Lelong, Journée d'Automne, Soc. Francaise Métal, (1957).

49. T. Sheppard and M.G. Tutcher, Met. Tech., 8 (1981), pp. 319-27.

50. E. Evangelista, H.J. McQueen and E. Bonetti, Deformation of Multiphase and Particle Containing Materials, ed. J. Bilde-Sorensen, RISO Natl. Lab. Roskilde, Denmark (1983), pp. 243-50.

51. A.R. Jones, B. Ralph and N. Hansen, Met. Sci., 13 (1979), pp. 149-54.

52. W.D. Donnelly and M.L. Rudie, Trans. Met. Soc. AIME, 230 (1964), p. 1481.

53. I. Miki and H. Warliment, Z. Metallkde, 59 (1968), pp. 254-64, pp. 408-14.

54. K. Holme and E. Hornbogen, J. Mat. Sci., 5 (1970), pp. 655-68.

55. R.H. Bush, Scripta Met., 1 (1967), pp. 75-78.

56. A.R. Jones and N. Hansen, Recryst. and Grain Growth of Multiphase and Particle Containing Materials, RISO Natl. Lab., Roskilde, Denmark (1980), pp. 13-25.

57. J.L. Brimhall, M.J. Klein and R.A. Huggins, Acta Met., 14 (1966), pp. 459-66.

58. J.B. Bilde-Sorensen, Deformation of Multiphase and Particle Containing Materials, RISO Natl. Lab., Roskilde, Denmark (1983), pp. 1-14.

59. J.M. Oblak and W.A. Owczarski, Met. Trans., 3 (1972), pp. 617-26.

60. L. Lee, Influence of Thermal History on Properties of Hydrostatically Extruded Aℓ-Fe alloys, M.S. Thesis, Stanford University, Palo Alto (1978).

61. G. Gottstein and U.F. Kocks, Acta Met., 31 (1983), pp. 175-88.

62. T. Hasegawa, T. Yajou and U.F. Kocks, Acta Met., 30 (1982), pp. 235-43.

63. R.A. Petkovic, M.J. Luton and J.J. Jonas, Can. Met. Q., 14 (1975), pp. 137-45.

64. J. Hausselt and W. Blum, Acta Met., 24 (1976), pp. 1027-39.

65. J. Schey, Acta. Tech. (Budapest), 16 (1957), pp. 131-52.

66. R.W. Westerlund, Met. Trans., 5A (1974), pp. 667-72.

67. H.A. Lipsitt and C.M. Sargent, Proc. 2nd Int. Conf. on Strength of Metals and Alloys, ASM, Metals Park, (1970), Vol. 3, pp. 937-40.

68. D. Webster, R.F. Karlak and A.E. Vidoz, Microstructure and Design of Alloys. (ICSMA-3) Inst. Metals, ISI London (1973) Vol. 1, pp. 325-30.

69. G. Beghi, M. Grin and G. Piatti, Mem. Sci. Rev. Mét., 64 (1967), pp.

70. N. Ryum, J. Inst. Met., 94 (1966), pp. 191-2.

71. P.L. Morris and M.D. Ball, Recryst. and Grain Growth of Multiphase and Particle Containing Materials. RISO Natl. Lab., Roskilde, Denmark (1980), pp. 97-102.

72. J.P. Lyle and R.J. Towner, U.S. Pat. No. 2, 963, 780, 13 December (1960).

73. U. Koster, Metal Sci., 8 (1974), pp. 151-60.

74. R.D. Doherty, Recryst. and Grain Growth of Multiphase and Particle Containing Materials, RISO Natl. Lab., Roskilde, Denmark (1980), pp. 57-70.

75. N. Hansen and B. Bay, J. Mat. Sc., 7 (1972), pp. 1351-62.

76. F.J. Humphreys and J.W. Martin, Acta Met., 14 (1966), pp. 775-81.

77. R.D. Doherty and J.W. Martin, J. Inst. Met., 91 (1963), pp. 332-38.

78. P.R. Mould and P. Cotterill, J. Mat. Sci., 2 (1967), pp. 241-55.

79. D.T. Gawn and G.T. Higgins, Textures in Research and Practice, Springer-Verlag, Berlin (1969), p. 319.

80. E. Nes, Met. Sci., 13 (1979), pp. 211-15.

81. E. Nes, Acta Met., 24 (1976), pp. 391-98.

82. E. Nes, Scripta Met., 10 (1976), pp. 1025-28.

83. E. Nes, Recryst. and Grain Growth of Multiphase and Particle Containing Materials, RISO Natl. Lab., Roskilde, Denmark (1980), pp. 85-95.

84. I. Baker and J.W. Martin, Strength of Metals and Alloys (ICSMA-6), ed. R.C Gifkins, Pergamon Press, Oxford (1982), Vol. 1, pp. 487-92.

85. I. Weiss and J.J. Jonas, Met. Trans., 10A (1979), pp. 831-40.

86. K.J. Gardner and R. Grimes. Met. Sci., 13 (1979), pp. 216-22.

87. B.M. Watts, M.J. Stowell, B.L. Baikie and D.G.E. Owen, Met. Sci., 10 (1976), pp. 189-97.

88. J.C. Blade, Met. Sci., 13 (1979), pp. 206-10.

89. P. Furrer and G. Hausch, Met. Sci., 13 (1979), pp. 155-62.

90. C.A. Romanowski and P. Cotterill, Recryst. and Grain Growth of Multiphase and Particle Containing Materials, RISO Natl. Lab., Roskilde, Denmark, (1980), pp. 127-32.

Acknowledgements

The authors are indebted to Paul McQueen for preparation of the graphs, to Janet Bowles for page layout and Jayne Claassen for word processsing.

STRUCTURAL EVOLUTION DURING THE ROLLING OF ALUMINIUM ALLOYS

T. Sheppard, M.A. Zaidi, P.A. Hollinshead, N. Raghunathan

Department of Metallurgy and Materials Science,
Imperial College of Science and Technology,
London, S.W.7.

Summary

The development of a substructure when aluminium is rolled has been investigated by TEM examination of material extracted from the roll bite using a quick release mechanism. It is shown that deformation occurs before the observation of macroscopic flow and that the subgrain size stabilizes soon after entering the roll bite. Results of multi-pass rolling experiments indicate that it is the last pass temperature compensated strain rate Z which determines the substructural topology. The experiments also suggest that the mechanism of subgrain formation involves migration of sub-boundaries. Subgrain growth is shown to occur between rolling passes and to be dependent upon the processing conditions. Recrystallisaticn after deformation is investigated and formulae predicting this behaviour are presented. It is shown that varying the hot rolling parameters influences the texture developed and that this can produce differences in earing in material which has been subsequently annealed and cold rolled.

Introduction

Considerable tonnages of Aluminium Alloys are processed and used in the form of strip sheet and plate. The hot rolling schedule is constructed to reduce slabs of up to 450mm thickness into thicknesses varying from about 3mm to 75mm. Whether subsequently cold rolled, heat treated or utilised in the hot rolled condition, the properties of aluminium alloys are influenced by the characteristics of the microstructure produced during the hot rolling process. For example, the substructure produced in the hot rolling process can affect the earing properties after considerable cold work (1), substructural strengthening can be an important feature (2) and recrystallisation and grain growth can persist through several passes having a deleterious effect on properties (3).

It is generally recognised that most dilute aluminium alloys recover to a subgrain structure during hot working (4) (5) and it has been accepted (6) that this structure is possible by a mechanism which involves the repeated unravelling and forming of the subgrain boundaries. It has been shown (7) that the coalescence of these subgrains plays an important role in subsequent recrystallisation processes.

The work presented in this communication, performed over a number of years, illustrates the importance of the development and morphology of the substructure developed during single and multi-pass rolling, discusses the mechanism of substructural change and investigates the effect of substructure on some important properties.

Experimental

The materials used were supplied by Alcan International and Alcoa (USA) in the form of DC cast logs and were subsequently machined into 200mm long billets of a suitable cross section for laboratory rolling. The composition of the alloys reported in this work is given in Table I.

Table I.   Composition of Experimental Alloys.

Designation	Si	Mn	Cr	Cu	Ti	Fe	Mg	Zn	B	Al
AA 1100	0.04	0.002	–	0.001	0.012	0.15	0.001	0.012	.0020	Bal
AA 3004	0.30	1.32	–	0.27	–	0.62	1.15	0.22	–	Bal
AA 5052	0.17	0.26	–	0.01	0.01	0.3	2.0	0.01	.0012	Bal
AA 5056	0.05	0.13	0.12	0.003	0.014	0.16	4.98	0.01	.0008	Bal
AA 5083	0.05	0.70	0.17	0.002	0.016	0.18	4.55	0.01	.0014	Bal

Most of the experimental details have been previously reported (8), (9), (10) and will not be elaborated in this communication. Where relevant, they are introduced into the text.

Results and Discussion

Development of macrostructure (experimental alloy AA 1150)

Fig.1 shows the deformation zone of a 25mm x 25mm slab rolled at 500 C with a strain rate of 2 s^{-1} and a 40% reduction. Deformation is inhomogeneous, initially perpendicular planes being bent backwards at entry to the roll gap and extended forwards at the midplane at exit to the roll gap. It is interesting to note that considerable deformation occurs before the roll bite, especially towards the centre of the slab. At the roll surface/material interface there is evidence of shear, and it appears that sticking friction prevails from entry to exit of the roll gap. There is a considerable temperature difference between the rolls and the material, and hence the surface layers are quenched, leading to larger flow stress values there than in the centre of the slab. This effect combined with friction would appear to lead to greater backward extrusion from the roll gap at entry, and to planes being heavily distorted at exit. Although the impression is given that the deformation is more severe at the centre, this is not necessarily the case because the original grains have been reorientated at the centre (partly due to spread) but not at the surface where friction constraints have maintained the original direction. Fig.1 also indicates the positions from which the specimens for electron microscopic examination were extracted.

Development of substructure (experimental alloys AA 1100 and 5052)

The development of the substructure as material passes through the roll gap is shown in Figs. 2 - 5. Fig.2 is taken from a foil extracted from a position 8mm ahead of the roll bite, indicated by A in Fig.1. It shows a triple junction of high misorientation containing α -(Al-Fe-Si) particles aligned coincident with the grain boundaries. There is no evidence of subgrains, the only substructural feature being dislocation lines which terminate either at particles or at grain boundaries. This is typical of a cast and homogenised structure. At a position closer to the roll gap (location B in Fig.1) but still generally considered to be outside the quasi-static deformation zone, an imperfect cellular structure with ragged walls and considerable dislocation activity may be observed (Fig.3). It is perhaps surprising that this evidence of deformation can clearly be seen to have originated in a region where apparently no macroscopic deformation has occurred. It indicates that very low strains are required to produce sufficient activity to initiate the formation of subgrains. Fig.3 shows that many dislocation tangles have formed in the grain interior but that well defined sub-boundaries (arrowed) form only at the grain boundaries and at those positions where larger particles are located. This is because the grain boundaries are likely to be locations of greatest stress in lightly deformed material. At these positions, larger strains are required to maintain continuity with adjacent grains. The particles also act as obstacles to dislocation motion because internal stress fields are created when there is a mismatch between the parent lattice and impurity atoms. At the onset of observable macroscopic deformation (position C in Fig.1, 1 mm into the roll bite) the electron micrograph (Fig.4) indicates a structure consisting of well developed subgrain walls but containing a mutilplicity of 'microcells' within

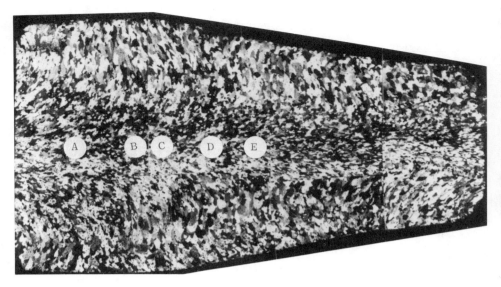

FIGURE 1. Deformation of
grains in roll gap;
billet entry
temperature = 500°C,
40% reduction, $\dot{\varepsilon} = 2\ s^{-1}$

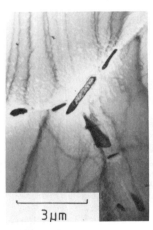

FIGURE 2. α(Al-Fe-Si)
particles aligned
along grain boundaries

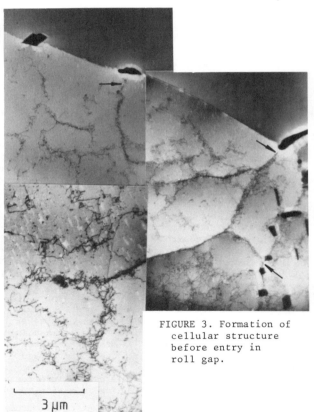

FIGURE 3. Formation of
cellular structure
before entry in
roll gap.

each subgrain.Aluminium is a high stacking fault energy (SFE) material
in which the relative rates of climb and the propensity for cross-slip
are high. Thus at low strains a uniform dislocation density is
unlikely to exist and the dislocations continuously assume positions
which form microcell boundaries by a combination of the processes
described above. Thermal fluctuation and the action of applied
stress unpin dislocations in the microcell boundaries which relocate
in more stable boundaries or interact with dislocations of opposite
sign and are annihilated. The effect of long-range stresses
associated with dislocations in one boundary also interacts with
dislocations in adjacent microcell boundaries, thus assisting in the
unpinning operation. Hence, because the applied stress at this
location is small, microcells are more likely to be observed.
Although this mechanism is presumably continuous (particularly at low
strains), the driving forces at higher applied stresses suggest that
microcell boundaries are less likely to remain stable. Within the
quasi-static deformation zone (position D in Fig.1) the subgrain walls
become better defined and the subgrain size decreases (Fig.5). In a
billet rolled at 500 C with 40% reduction, the average subgrain
diameter decreased from 9 to 6.4μm within the first 10mm of the
projected length of contact (position E in Fig.1). The total length of
contact for such a reduction is 35mm, and beyond 10mm the subgrain
size is stabilized and no further change in size could be observed.
At this stage the interior of the subgrain has a clean appearance with
low dislocation density because the well developed subgrain walls are
very effective sinks for further generated dislocations which interact
with dislocations of opposite sign within the walls and thus are
mutually annihilated. The stabilization of the subgrain size early
in the roll gap could have important practical implications. It
illustrates that the use of a mean strain rate and an average
temperature rise is unlikely to result in the calculation of a
temperature compensated strain rate or a Zener-Hollomon parameter Z
consistent with the final substructure. In the given example the
temperature compensated strain rate at the point of stabilization can
be shown to be Z = 1.52 x 10^{11} s^{-1}, while the average for the
total pass is Z = 8 x 10^{10} s^{-1}.

Subgrains will also form in solid solution alloys (e.g. Al-2Mg)
by the mechanisms discussed above, but differences in subgrain size
and substructure features can be recognised. The activation energy
for interdiffusion of magnesium in Al-Mg alloys is 135 kJ mol^{-1} and
for selfdiffusion is 156 kJ mol^{-1} in the hot working range.
Magnesium atoms therefore diffuse more rapidly and surround
dislocations, thus providing more obstacles to dislocation activity in
the Al-Mg alloy than would be expected in the commercially pure alloy.
The net result is that more boundaries may be expected, producing a
smaller subgrain size when processed identically. Electron
micrographs of the two alloys produced at Z = 0.2 x 10^{11} s^{-1} are
presented in Figs. 6a and 6b. The subgrain size is clearly smaller
in the AA 5052 alloy, and sub-boundaries are less well formed, and the
dislocation density within the subgrains is high. It is well known
that there is strong binding between vacancies and the Mg atoms in
Al-Mg alloys. The process of climb, being dependent on vacancy
availability, is impeded, and these two factors would appear to be the
reasons for the observation of ill formed boundaries and high
dislocation densities in the subgrain interior throughout the Al-Mg
hot-working range.

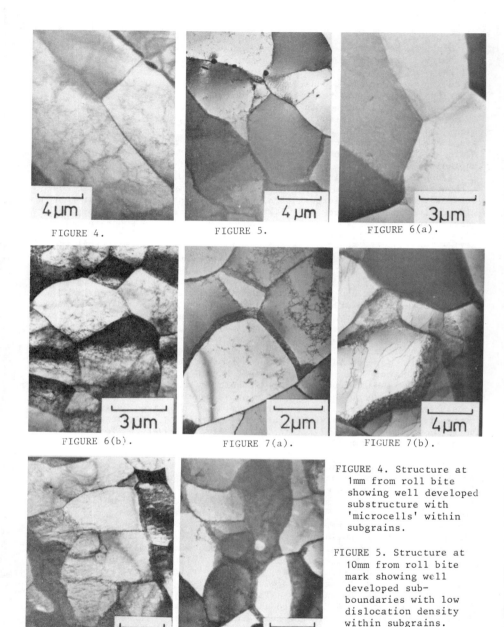

FIGURE 4.

FIGURE 5.

FIGURE 6(a).

FIGURE 6(b).

FIGURE 7(a).

FIGURE 7(b).

FIGURE 7(c).

FIGURE 7(d).

FIGURE 4. Structure at 1mm from roll bite showing well developed substructure with 'microcells' within subgrains.

FIGURE 5. Structure at 10mm from roll bite mark showing well developed sub-boundaries with low dislocation density within subgrains.

FIGURE 6.(a) low Z substructure, alloy 1S; (b) low Z substructure, alloy M57S.

FIGURE 7.Substructure of billet formed after various rolling schedules; specimens held for 30 s after final pass before quenching; $Z = 3 \times 10^{12} s^{-1}$ (a) single pass; (b) two passes; (c) three passes; (d) four passes.

Multipass Rolling (Experimental Alloy AA 1100)

Details of the rolling schedule utilised during these experiments are given in Table 2.

Table 2.   Rolling Schedule.

Pass	Reduction per pass, mm	Pass geometry Reduction %	Strain	Strain rate s^{-1}	Temperature before entering next pass °C	Rest period before entering next pass °C
1	5	20	0.223	2.0	460	30
2	5	25	0.228	2.57	400	30
3	5	33.3	0.405	3.62	345	30
4	5	50	0.693	6.2	-	-

Typical electron micrographs of the substructures formed during the multipass rolling schedules are shown in Fig.7. Each specimen was held for 30s after completion of the final pass before being quenched. There is nothing in the figures to suggest that the multiple pass had any significant effect. The subgrains are well formed in each case, and the variation in size (decreasing with each pass) would be expected because the temperature of deformation also decreases. The structure is perhaps more equiaxed in the single-pass sequence, and there is also evidence of microcells within subgrains in this illustration (Fig.7). These microcells are associated with the early stages of deformation, but are also more frequently observed at higher temperatures. Fig.8 shows a micrograph from a specimen deformed from 50 to 25mm in a single pass. The subgrain size and morphology are almost identical to that shown in Fig.7 indicating that it is the final pass which determines substructure.

The variation in subgrain size across the thickness of the billets is shown in Fig.9. The large variation in the one-and-two-pass rolled billets arises from the larger temperature variation (11) from surface to centre in the early passes; there is also a greater inhomogeneity of deformation, the centre of the billet undergoing considerably less strain. The strain in each pass was 0.2 which, although typical of an early pass in a rolling schedule, is still insufficient to provide appreciable deformation penetration at the centreline of the specimen in the first pass. Similar conditions exist during the initial passes in industrial rolling schedules, when the billet dimensions may typically be 250-350mm thick and 1000-1250mm wide: certainly not plane-strain conditions. The effect of decreasing thickness in providing greater temperature and deformation homogeneity is illustrated in Fig.9 which shows the subgrain size becoming more uniform until, after four passes, the subgrain size variation is less than 1μm. The variations shown in Fig.4 may have important industrial implications because variations in through-thickness subgrain size might still be recognizable in the thicker plate gauges.

In the specimens allowed to cool for 30s before quenching, a signficant number of subgrains in the centre of the billet were found

25

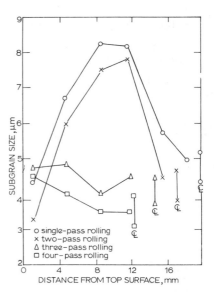

← FIGURE 8.
    Microstructure
    of billet
    reduced by 50%
    during single
    pass and held
    for 30s before
    quenching.

FIGURE 9. →
    Variation of
    subgrain size
    across thick-
    ness of multi-
    pass-rolled
    billet.

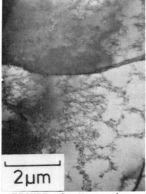

FIGURE 10. Formation of
    microcells within sub-
    grain (upper subgrain),
    together with fully
    developed microcells
    (lower subgrain).

FIGURE 11.
    Deformation bands
    (a) low misorientation
    (b) high (12°)
        misorientation.

FIGURE 12.
    Structure of
    specimen about to
    enter final pass
    of four-pass
    schedule, showing
    deformation bands
    in all grains.

to contain microcells (Figs.7a and 10). The microcells were similar in appearance to those which have been reported to precede steady-state subgrain formation. It is clear that such microcells can form after deformation, some by climb from the deformation-induced sub-boundaries, but most by dislocation climb from within the as-deformed subgrains. The rearrangement of randomly distributed dislocations produce both annihilation and reductions in the magnitude of long-range stress fields. Hence the observation of such microcells is indicative that static recovery occurs during the interpass time. So, whenever sufficient thermal energy is available, dislocations will arrange themselves into microcell boundaries. At lower temperatures randomly distributed dislocations were observed, but the formation of microcells was not evident.

Deformation bands were observed in each of the rolled billets and were readily recognizable; both in the optical and the electron microscope. Although observed by several workers (12) (13) the mechanisms by which deformation bands are formed is still not clear. Rhines (14) and some of his co-workers have suggested that the phenomenon becomes apparent when deformation of a pair of grains is possible in one direction only. To accommodate a shape change it is necessary for one of the grains to deform by the alternate tilting of bands running across the crystal. Dillamore et al (15) suggested that stress variations within a crystal could cause relative rotations between adjacent segments of the grain. Ahlborn (16) has proposed that deformation bands are created wherever local differences in stress state activate alternative combinations of slip systems in order to produce the same imposed microstrain at specific locations in the crystal. Different slip rotations will then occur, so producing relative misorientations. In general deformation banding is not observed, or certainly is not a prominent feature, in fine-grained materials. In such materials extra slip systems are able to operate in the vicinity of grain boundaries and thus accommodate grain-boundary mismatch. Conversely, in larger-grained material (as used in the present work) deformation bands may form very early in the deformation process (Fig.11 exhibiting low misorientation across the band boundaries $(1.5^{\circ} - 2^{\circ})$. With increasing deformation, the misorientation across the band increases to $10^{\circ} - 15^{\circ}$ (Fig.11), and bands are more likely to form. The presence of large misorientations at the deformation-band boundaries can be an important factor in promoting the growth rate of nuclei during the recrystallisation process (17).

Effect of Prior Substructure on Subgrain Formation

Theoretically, the developed subgrain size should depend on the values of the deformation variables at the final pass of a given multipass schedule; the subgrain size having been generally recognised to be independent of deformation history. However, during multipass rolling, the structure is required to develop from quite different entry structures. To investigate the effect of prior substructure and the mechanisms by which changes in subgrain size are effected, the rolls were sprung apart during the final pass of a four-pass rolling schedule. Details of the schedule are given in Table 2. A micrograph showing the structure of typical material entering the final pass is shown in Fig.12. The structure is unrecrystallised, with deformation bands clearly visible in most of the grains. Other pass schedules employed in the experimental programme did show evidence of recrystallisation, and this is discussed below.

It has been reported (6) that, for subgrains to change size, it is necessary for the existing boundaries to unravel completely before the new boundary can form; observations possible from the results of these experiments offered an alternative explanation. During subsequent rolling the subgrains established by the previous pass were observed to 'bend' or 'fragment', appearing in two dimensions as a division of subgrains. The 'subdivided' boundaries (arrowed in Fig.13) then migrate to form subgrains of a size commensurate with the strain-rate and temperature conditions obtaining in the deformation zone. The boundaries of the subdivisions were also observed to be weak, and showed a marked tendency to migrate in the direction of larger dislocation densities; they may contribute to recovery by dislocation annihilation, as suggested by Exell and Warrington (18). These workers also observed that, under the influence of low-magnitude stresses (10 MNm^{-2}) at high temperatures (400°C - 600°C) sub-boundaries migrated at high speed (10µm s^{-1}). Hence, at the temperature-compensated strain rates and the stresses encountered in hot rolling, the rearrangements required to form an 'equilibrium' subgrain size would appear to occur very rapidly. Hence the complete disintegration of sub-boundaries, as suggested by McQueen et al (6) need not be a necessary condition for deformation. The probable reaction of a substructure to further deformation would appear to commence by dislocations separating from sub-boundaries and weaker links surrounding particles, etc. Such dislocations interact to form tangles which migrate under processing conditions until a subgrain attains its equilibrium size. This is clearly illustrated in Fig.13 where the 'subdivided' subgrains A - D are of approximately the same size as the well-formed subgrains E and F.

Another mechanism of sub-boundary formation in a prior substructure is when unpinning of dislocations occurs because of the action of a pile-up on a weakly formed boundary. These unpinned dislocations may then glide, interact with other dislocations, and form a new boundary. Such a process was observed by depositing a carbon film on a foil by electron beam heating (in the absence of an anticontaminator) and observing the dislocation activity generated by the stresses developed due to the deposited carbon film. The experiment was performed at room temperature, but since the substructure is characteristic of hot working it is probable that such activity could also occur at higher temperatures. Under the action of stress, dislocations were observed to be generated at sources in the sub-boundaries. A surge of such generations provided a stress concentration sufficient to disintegrate part of the boundary and form a bulge (Fig.14a and b). The unpinned dislocations interact with other dislocations to form a new boundary (Fig.14c - f). Fig.14 shows the action under conditions of very low stress and so the boundaries developed are not well defined, but at higher stresses and temperatures this would appear to be a possible mechanism of sub-boundary formation.

Subgrain Growth during Interpass Time

The subgrain size was measured in specimens quenched immediately after the final pass and compared with those measured in specimens subjected to a 30s delay before quenching. It was found that

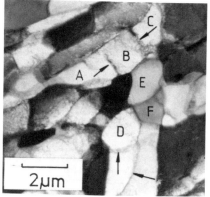

FIGURE 13. Weakly formed sub-boundaries
(arrowed) inside well formed subgrains;
sub-divided subgrains A-D are similar
in size to well formed subgrains E and F.

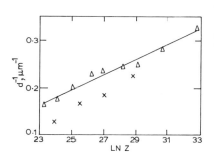

FIGURE 15. Effect of Z on
developed subgrain size
(Δ) and subgrain growth
(x) between passes.

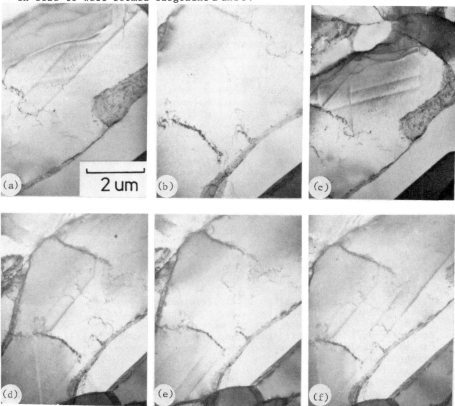

FIGURE 14. Disintegration of part of subgrain boundary and formation
of new boundary.

considerable subgrain growth had occurred during the delay. Fig.15 shows a comparison of the sizes evaluated for the same temperature-compensated strain rate Z and it can be seen that the subgrain size increases by about 35% at the lower Z (2×10^{10} s^{-1}) and 22% at the higher Z (3×10^{13} s^{-1}). In the hot worked structures investigated, it has been established (7) that subgrain coalescence is the mechanism by which nucleation of recrystallisation occurs. It is also clear that the extent of subgrain growth is determined by the subgrain size in the previous pass, and the interpass time.

Recrystallisation between Passes

The single- and two-pass-rolled billets having an interpass time of 30s were observed to be partially recrystallised. Fig.16 shows typical structures of the two billets. The optical micrograph taken after one pass [Fig.16] shows clearly the growth of a recrystallised grain into a banded region, while the micrograph taken after two passes [Fig.16] suggests that nucleation and growth occurs mainly at grain boundaries. Each of these sites, in any case, will be competing sites for nucleation, and it is possible that the etching technique has not revealed the deformation band. There seems no reason to suppose that the preferential nucleation should change after single-pass rolling. The observation is important because it is not uncommon for industrial rolling schedules to include passes under conditions similar to the first two passes utilized in this experimental sequence. Clearly, if recrystallisation occurs during the early passes in a deformation schedule, then those passes cannot be additive in the 'texture' sense. It should be stated, however, that the relationship between total hot-working strain and development of texture is obscure and probably will not be as pronounced as that found during the cold working of metals.

The deformation bands were observed frequently in the rolled material, and a mechanism by which they can form has been described above. In all the billets which had been reduced by more than 40%, each grain was observed to contain such bands. This is illustrated clearly in the four-pass rolling schedule billet shown in Fig.16 and it is the high misorientations across the band ($10°$-$15°$) which can greatly enhance the growth rate of nuclei. Subgrains grow until they impinge on a deformation-band boundary, thus acquiring a partial high-angle boundary capable of growth into the deformed matrix. Thus, nuclei observed within grains were invariably associated with grains containing deformation bands.

Recrystallisation after final deformation was not observed in those billets which received three or four passes. Most of the grains had a heavily banded appearance. Note that the exit temperature of the third pass was approximately $400°C$, and hence it can be concluded that recrystallisation does not occur for values of Z (last-pass) greater than 2×10^{12} s^{-1}, for this alloy and for deforamtions <20% per pass. Most of the recrystallised grains were smaller than the original grains, but there was evidence of excessive grain growth having occurred, especially on those billets rolled in one pass at 525°C, held for 30s, and quenched. The larger grains were usually located at the mid-section of the billet (Fig.17), but it should be borne in mind that in the industrial process a much greater

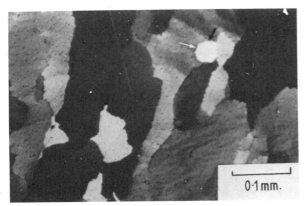

(a)

FIGURE 16.
Recrystallisation during 30 s between final pass and quenching.

(a) after one pass, showing growth of recrystallised grain (arrowed) into banded region;
(b) after two passes, showing nucleation and growth of recrystallised region.

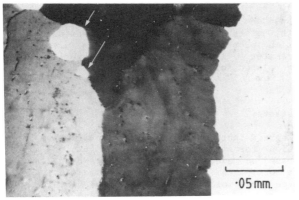

(b)

FIGURE 17. Very large recrystallised grain at mid-section of billet rolled in single pass at 525°C and quenched after 30 s.

proportion of the total thickness will remain at high temperature. Thus events occurring during experimental rolling may be more widespread in the practical case. These large grains are probably due to the growth of nuclei favourably orientated with respect to the deformation matrix. Such grains can have deleterious effects on properties, especially on texture formation and the surface quality of the final product. These very coarse grains can persist through the hot-rolling schedule, and during subsequent annealing and cold rolling, and if present at the surface give rise to the 'orange-peel' effect in the finished sheet.

In concluding this section it is essential to note that the dominant mode of recrystallisation was by nucleation within the grain interiors. The nucleation process is thus more closely connected with subgrains and their characteristics than with the original grain boundary morphology. Subgrains appear to grow until the formation of a high-angle boundary which is capable of sustained growth until it impinges on some other boundary or is arrested at a precipitate. The exact nature of this nucleation has been reported. (7).

Effect of Process Parameters on Subgrain Size

The temperature compensated strain rate or Zener Hollomon parameter Z incorporates both the strain rate and the temperature T prevailing during a deformation process; the flow stress is also a strong function of this parameter. It is therefore not surprising that the subgrain size may be related to Z in the form

$$d^{-m} = a + b \ln Z \tag{1}$$

where and a, b and m are constants.

Most workers (19) have agreed that making m = 1 in equation (1) provides the best correlation for aluminium and its alloys. A correlation was attempted based on equation (1) for the two alloys investigated in the present work: the billets being rolled in one pass and quenched. A large range of Z values was employed, and linear regression revealed that good fits could be obtained for m values varying from 0.3 to 1.25 (the lowest correlation coefficient was 0.94). Achieving such good fits is not really surprising because the range of subgrain sizes generally obtainable in the hot-working range is small when compared to the range of ln Z values: mathematically, the result would be expected. Assuming for the time being that a and b have no physical meaning, it would appear that any value of m between 0.3 and 1.25 gives a satisfactory correlation coefficient.

Substructure Strengthening

The mechanism of substructure strengthening has been discussed by many workers (20) (24) and it is generally accepted that a modified Hall-Petch relationship of the type

$$\sigma = \sigma_o + k_s d^{-m} \tag{2}$$

is valid for substructural barriers. It is not clear, however, whether the subgrain boundaries act as barriers to slip in a similar way to high-angle boundaries. The ultimate or proof stresses (i.e.

32

the flow stress) may be influenced by the nature of the subgrain walls as well as by the scale of the substructure. The spacing and geometry of dislocation networks in the subgrain wall will determine the magnitude and distribution of the stress fields surrounding them. In the model proposed by equation (2) σ_0 is invariant, which does not allow the stress fields at sub-boundaries to change. Most data fitted to the modified Hall-Petch relationship are based on 'creep subgrains' (7) or subgrains produced by recovery annealing (22) of cold-worked substructures. These subgrains normally have low internal dislocation densities and few redundant dislocations in the boundary. They therefore tend to act as barriers of constant strength (i.e. constant k_s) and hence a good fit is usually obtained using equation (2). In hot-rolled material the subgrain interior dislocation density is often large, especially when the subgrain size is small or the alloy is one which contains many second-phase particles and precipitates. The dislocation tangles and microcells described above also act as strengthening features. Abson and Jonas (23) also suggested that the sub-boundaries increased in strength with decreasing subgrain size.

Clearly, any attempt to fit equation (2) must be approached with some caution. In general, a value of m = 1 has been accepted as fitting the data for subgrain strengthening (20). However, in most works reported, the value of σ_0 appears to be anomalously low, and sometimes a negative value has been quoted (24). This is meaningless since σ_0 represents the strength of a subgrain-free material. These low values of σ_0 have probably been obtained because the strengthening effect of subgrains increases non-linearly as the spacing decreases because of the changes in the character of the sub-boundary and subgrain interior. The subgrain size is normally small (< 5μm) in the hot-working range of commercial alloys; hence a large variation of m values could be expected to give a reasonable fit when plotted against the much larger variation in σ. Thus the value of m should not only produce a good fit to equation (2) but should ensure that the intercept σ_0 is close to the strength of the annealed material. Various m values were inserted into equation (2) for the two alloys. The strength of annealed commercial purity aluminium is 37.8 N mm^{-2} and there m = 3 would be appropriate for this alloy. Similarly, m = 1.35 gives a reasonable value for the annealed strength of the AA 5052 alloy. The resulting equations for the two alloys are :

$$\sigma = 37.8 + 1856d^{-3.13}$$

for commercial purity AA 1100 and

$$\sigma = 78.68 + 59.26d^{-1.35}$$

for alloy AA 5052, which indicate sub-boundary strengthening to be greater in the commercial purity 1S. This would be expected since the boundaries in AA 5052 are less well defined and so it should be easier to unpin a dislocation from the boundary of such an alloy. The hot work induced subgrain size depends upon the deformation variables [equation (1)] prevailing during the rolling process, which implies that a linear relationship should exist between the room-temperature strength and the deformation variables defined by Z. Fig.18 shows the relationship for the alloys investigated and it can be seen that linear relationships do exist. Such data could be used in a predictive manner to assist in structure control during rolling.

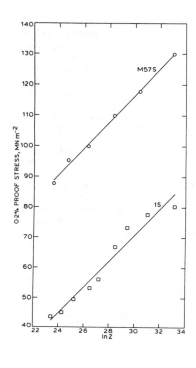

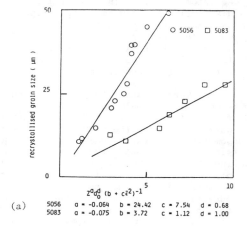

FIGURE 18. Effect of hot-rolling process parameters (defined in Z) on room-temperature proof stress.

(a)

5056	a = -0.064	b = 24.42	c = 7.54	d = 0.68
5083	a = -0.075	b = 3.72	c = 1.12	d = 1.00

FIGURE 19.
(a) Effect of processing on recrystallised grain size.

(b) Complex function determines time for recrystallisation

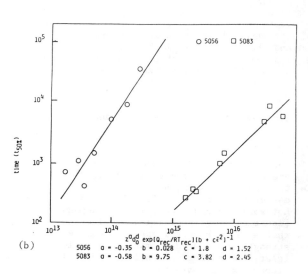

(b)

5056	a = -0.35	b = 0.028	c = 1.8	d = 1.52
5083	a = -0.58	b = 9.75	c = 3.82	d = 2.45

Effect of Process Parameters on Recrystallisation
Experimental Alloys AA 5056, AA 5083

The discussion above has revealed that recrystallisation between passes and the processing parameters during a pass are related parameters. This is primarily because the mechanism of nucleation is by subgrain coalescence. However since nucleation does occur at grain boundaries and is itself a thermally activated process we may expect the time for recrystallisation and the grain size to be a function of many parameters. Consider first the well known Avrami equation which is valid for isothermal conditions and a singular nucleation mechanism :

$$x_v = 1 - \exp(-Bt_x^n) \qquad (3)$$

where x_v is the fraction recrystallised in time, t_x and B and n are constants. The temperature dependence of static recrystallisation may be allowed for (25) by introducing the concept of a temperature compensated time W_x such that

$$W_x = t_x \exp\left(\frac{-\Delta H_{rex}}{GT_A}\right) \qquad (4)$$

where ΔH_{rex} is the required activation energy for recrystallisation, G is the universal gas constant and T_A is the annealing temperature. But the time t_x for recrystallisation of a of a given fraction X_v is also dependent upon the subgrain size which is a function of the temperature compensated strain rate Z, the total hot strain representing the extension of the original grain surface area, the intial grain size d_o and possibly the spacing of large particles . We might, therefore, expect that the form of the equation for the term required to recrystallise $X_v\%$ would be :-

$$t_x = k.\lambda^a.d_o^b.\,Z^c.\varepsilon^d \left(\exp\frac{\Delta H_{rex}}{GT_A}\right) \qquad (5)$$

An important feature of a microstructure is the recrystallised grain size which we would also expect to be dependent on the density of nucleation sites. The subgrain size after the final rolling pass should therefore be an important parameter, governing the subgrain coalescence mechanism and being related to the temperature compensated strain rate obtaining during the final pass. The density of nucleation sites should also be related to the grain boundary area which will be a function of total strain and initial grain size if interpass recrystallisation is prevented. Coarse particles may again effect recrystallised grain size suggesting that an appropriate form of equation should be :-

$$d_{rex} = K_1.d_o^p.\,\varepsilon^q.\,Z^r.\,\lambda^s \qquad (6)$$

The experimental validation of equation (5) and (6) requires much careful and painstaking work. The Avrami equation must first be satisfied, the activation energy for recrystallisation established and each variable investigated independently. Rolling experiments were performed on alloys 5056 and 5083 in which the temperature compensated strain rate Z was varied between 4×10^{12} and 3.5×10^{19}, the strain between 0.4 and 2.69 and the grain size between 150μm and 170μm. It was not possible to vary the interparticle spacing. The results for recrystallised grain size are those for 50% recrystallisation are shown in Figs. 19 and 20 respectively. It was found in each case that the strain could be better incorporated into the equations by writing $\dot{\varepsilon} = (B + C)^n$ such that the resultant equations could be written

For AA5056 Alloy $t_{50} = 9.1 \times 10^{-12} d_o^{1.52} Z^{-0.35} (0.0286 + 1.8\varepsilon^2)^{-1} \exp(\frac{212000}{GT_A})$

For AA5083 Alloy $t_{50} = 2.7 \times 10^{-10} d_o^{2.45} Z^{-0.58} (9.75 + 3.82\varepsilon^2)^{-1} \exp(\frac{183000}{GT_A})$

For 5056 Alloy $D_{rex} = 101.72 d_o^{0.68} Z^{-0.064} (24.42 + 7.54\varepsilon^2)^{-1}$

For 5083 Alloy $D_{rex} = 4.79 d_o \quad Z^{-0.075} (3.72 + 1.12\varepsilon^2)^{-1}$

Thus the recrystallised grain size in the F temper can be seen to be very sensitive to the hot rolling process parameters and consequently the cold rolling process will inherit differing structures from the prior deformation. Cold rolled properties can thus be expected to vary with hot rolling conditions.

Texture Development during Hot Rolling
(Experimental Alloy AA 3004)

It has been established that Aluminium, when dilutely alloyed, develops a rolling texture of the 'copper' or 'pure metal' type (26), this being primarily due to the materials high stacking fault energy. Such a texture may be described by a spread of orientations centered from {110} <112> through {427} <232> to {112} <111> as shown in Fig.21a.

Examination of full (100) and (200) pole figures produced from 3004 Alloy subjected to multi-pass hot rolling at temperatures between 250°C and 500°C indicates that the basic contours of the 'pure metal' type pole figures are well developed when the hot worked metal has received reductions of around 50% (Fig.21b). The pole figures obtained from material processed at differing temperatures appeared to be substantially invariant, suggesting that the strength of texture produced in the early passes of a rolling schedule is not affected by the rolling process conditions. The plot intensity levels of the pole figures clearly showed that, at equivalent strains, there was no systematic variation with temperature.

At lower rolling temperatures further rolling gradually sharpened the texture, and reductions greater than 80% produce fully developed 'pure metal' textures as illustrated in Fig.21c for deformation at room temperature. In general, such figures were produced for rolling temperatures up to 300°C. This gradual sharpening of texture does not seem to be achieved when processing at higher temperatures (Fig.21c) and appears to be caused by a very small volume fraction of recrystallised material which is produced during the quench-reheat interval of the pass schedule thus destroying part of the rolling texture component produced. This interpass recrystallisation occurs predominantly in the latter passes of the schedule because the effect of roll quenching is greater at this stage. In effect the temperature is reduced and the effective anneal temperature is thus greater than the rolling temperature. The energy available for

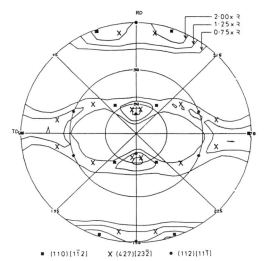

■ (110)[1̄1̄2]   X (427)[23̄2̄]   ● (112)[111̄]

[111] pole figure of AA 3004 90% reduced at room temperature

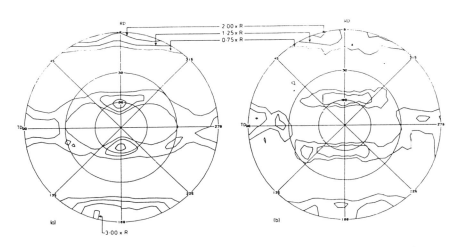

[111] pole figures of specimens reduced 83%. (a) at room temperature
(b) at 520°C.

FIGURE 20. Influence of hot strain and temperatures on pole figures.

FIGURE 21. Effect of temperature on the φ = 45° section of ODF

(a) as rolled

(b) annealed

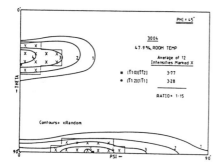

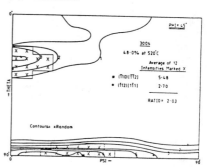

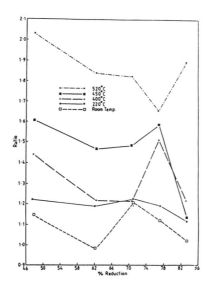

FIGURE 22. Ratio of average of 12 intensities about [$\bar{1}10$][$1\bar{1}2$]
and [$\bar{1}12$][$1\bar{1}1$] on φ = 45° section of ODF against reduction.

recrystallisation is also correspondingly greater, and recrystallisation occurs during the reheat for the next pass. Such conditions would not of course be encountered in the industrial process.

ODF analysis was performed on data from full (111) and (200) pole figures in an attempt to clarify whether the initial observation that no systematic variation with temperature was correct. The analysis utilised materials hot rolled to the same equivalent strain (48%). Surprisingly, the analysis revealed systematic variations with the processing conditions. The $\phi = 45$ section of the ODF (using Roe's notation) was constructed for each pass, Fig.22a, and the average intensity of 12 locations around (110)[112] and (112)[111] and the ratios between these intensities calculated.The results are presented in Fig.22b and clearly indicate that an increase in rolling temperature results in an increase in the strength of components around (110)[112] compared to those around (112)[111]. This variation was still evident in material rolled to higher strains but was less prominent probably because of interpass recrystallisation. This temperature dependence is presumably associated with the greater mobility of climb recovery processes as opposed to cross slip with increasing temperature (Fig.22a). The ODF analysis also indicated that small amounts of cube orientation could be detected after each pass at the higher rolling temperature. This cube component was not however strengthened by a recrystallisation anneal (fig. 23).

The experiments thus indicate that the temperature at which the rolling process is performed may produce differences in texture. These differences are difficult to detect by observation of normal pole figures and it is not possible to qualify their significance.

Relationship between Earing Behaviour and Processing Parameters (Experimental Alloy 5052) (Experimental Alloy AA 5052)

It has been established that after hot rolling and subsequent annealing, the structure of material presented to the cold mill will have varying texture: it may have substructure, or may be recrystallised but with a grain size determined by prior deformation history. When the material is cold rolled, a deformation texture will be re-imposed which itself comprises two major components, the R deformation texture and the so-called 'alloy-type' {110}<112> texture. This alloy-type component is generally found to be present in Al-Mg alloys following cold deformation (15) although Blade (27) has indicated that in the industrial case considerable deformation is required before a stable texture 'mix' is achieved. The application of a 60% reduction in the cold-rolling operation will thus produce a mixed cube and deformation texture and, although some cube-orientated grains will rotate (27) Blade considered that a substantial proportion will retain their original orientation.

5052 material from some of the experiments described above was subjected to cupping tests after annealing and cold rolling to 60% reduction (HAC) and after annealing subsequent to the cold rolling operation. Fig. 24 shows that appreciable 45 earing was observed in all the HAC material. When annealed, hot-rolled material of subgrain size less than 1.8 μm had a considerably reduced grain size. On subsequent rolling the tendency to form deformation bands would thus be much lower than for non-annealed material. The orientation

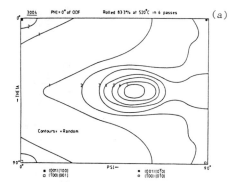
(a)

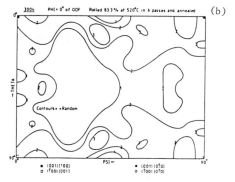
(b)

FIGURE 23. Effect of annealing on
  the ϕ = 0 section of ODF
  (a) as rolled
  (b) annealed

FIGURE 24. Earing behaviour
  of hot rolled annealed and
  cold rolled material.

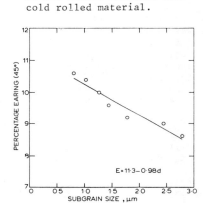

FIGURE 25. Earing behaviour of hot rolled,
  annealed, cold rolled (10%) and
  annealed AA 5052 alloy.

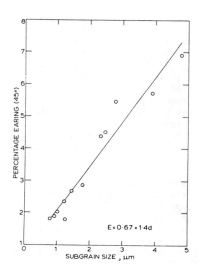

40

changes occurring during cold rolling would thus produce material closer to the stable end-orientation, manifesting itself in high earing levels. This suggests that if the grain size before hot rolling had been small, the earing produced after hot rolling and cold rolling would have been higher than was observed. It also suggests that since the as-cast grain size of this material was larger than that present in a typical commercial DC cast slab, the overall earing levels at each stage of deformation would, in the practical case, vary from those reported here. The important conclusion to be drawn from Fig. 24 is, nevertheless, that the final earing characteristics are strongly influenced by the hot-working process and vary consistently with the substructural features of the F temper material.

When the HACA material was deep drawn, ears were formed at 45° to the rolling direction in all cases. However, the trend was towards 90° earing with a decrease in the original hot-rolled subgrain size (fig. 25). Such behaviour may be explained in terms of the established texture developments during cold rolling and the reaction of this texture to an anneal.

First, the earing results indicate that the grain orientations in material presented to the cold mill depended on which hot-process parameters were used. Another important factor is that the different hot-worked substructures result in differing grain sizes being presented to the cold mill. The annealed grain size is retained in the HAC material, and it is well established (28) that the preferential growth of cube grains is favoured only after the new grains have grown sufficiently to be in contact with the matrix in at least two and preferably in all four matrix orientations (the symmetrical R components). It is clear that this critical size is smaller, and therefore that it is attained earlier in the annealing cycle if the grains representing the four matrix orientations are smaller. Consequently, the percentage cube texture for a given annealing cycle will increase with decreasing grain size. Hence the cube texture in the final annealing process would decrease with increasing hot-work subgrain size, showing increased 45 earing.

Practically, the most important conclusion is that the percentage earing may be varied in the HACA condition by suitable choice of the hot-rolling parameters. The relationship between hot-work induced subgrain size and earing behaviour is certainly an indirect relationship, but it is nevertheless of some consequence.

In the practical industrial case, sheets for deep-drawing operations are produced in either the HAC or the HACA form. The discussion above has indicated that for both these conditions in an Al-2Mg alloy, the percentage earing can be adjusted by controlling hot-rolling practices. For HAC material a larger subgrain size in the F temper is preferred, and may be produced by a suitable combination of low strain rates and high finishing temperatures. For material annealed after cold rolling, a smaller subgrain size should be the aim.

Conclusions

It has been shown that the development of a substructure during rolling commemces at grain boundaries and large particles apparently at locations not generally considered to be in the deformation zcne.

Deformation continues with grain boundaries being primary sources of dislocation generation, forming unstable microcells and finally a stable subgrain arrangement. Surprisingly this stable subgrain size remains unaltered after about one third of the deformation zone has been traversed. The subgrain size evolved during a pass and hence all properties are to some extent dependent upon the properties evolved during the final pass. It is shown that the strength in the hot rolled condition is directly related to these processing conditions for two alloys and that the hot rolled texture varies with processing conditions. Recrystallisation after hot rolling can also be expressed as a function of the total hot rolling process (including the effect of accumulated strain during multipass rolling). Earing properties after subsequent cold rolling and after cold rolling and annealing may also be related to the hot rolled substructure.

Acknowledgements

The work reported in this communication has been supported by ALCAN INTERNATIONAL, by ALCOA and by ARCO METALS. This support has been supplemented by discussions with various personnel from these institutions and the authors express their appreciation.

References

1.   T. Sheppard and M.A. Zaidi, Met.Tech., 1982, 9, 368.

2.   M.A. Zaidi and T. Sheppard, Met.Sci., 1982, 16, 229.

3.   R. Grimes, Met.Sci., 1974, 8, 176.

4.   J.J. Jonas, C.M. Sellars and W.J. McG. Tegart, Met.Rev., 1969, 14, 1.

5.   W. Wong, H.J. McQueen and J.J. Jonas, J.I.M., 1967, 95, 129.

6.   H.J. McQueen, W.A. Wong, J.J. Jonas, Can.J.Phys., 1967, 45, 1225.

7.   M. A. Zaidi and T. Sheppard, Met.Tech., 1984, 11, 313.

8.   D. S. Wright, Ph.D. Thesis, University of London, (Imperial College), 1978.

9.   M. A. Zaidi, Ph.D. Thesis, University of London (Imperial College), 1980.

10. P.A. Hollinshead, Ph.D. Thesis, in preparation.

11. T. Sheppard and D.S. Wright, Met.Tech. June 1979, 215.

12. T. Sheppard, M.G. Tutcher, H.M. Flower, Met.Sci., 1979, 13, 473.

13. C.S. Barrett and F.W. Steadman, Trans. AIME, 1942, 147, 57.

14. F.N. Rhines, 'Inhomogeneity of Plastic Deformation', Chap.10, 1973, Metals Park, Ohio, American Society for Metals.

15. I.C. Dillamore, P.L. Morris, C.J.E. Smith and W.B. Hutchinson, Proc. R. Soc., 1972, A329, 405.

16. H. Allborg : 'Recrystallisation, grain growth and textures', 374:1974 Metals Park, Ohio, American Society for Metals.

17. S.P. Bellier and R.D. Doherty, Acta Metall. 1977, 25, 531.

18. S.E. Exell and D.H. Warrington, Philos. Mag. 1972, 26, 1121.

19.   H.J. McQueen and J.J. Jonas : Treatise on Material Science and Technology', vol.6, 404, 1975, New York Academic Press.

20.   R.J. McIlray and Z.C. Szkopiak, Int.Met.Rev., 1972, 17, 175.

21.   J.C.M. Li, Trans. Met. Soc, AIME, 1963, 227, 239.

22.   C.J. Ball, Philos. Mag. 1957, 2, 1011.

23.   D.J. Abson and J.J. Jonas, Met.Sci.J., 1970, 4, 24.

24.   H.J.McQueen, Metall. Trans. 1977, 8A, 807.

25.   D.R. Barraclough and C.M. Sellars, Met.Sci., 1979, 13, 257.

26.   M. Hatherley and W.B. Hutchinson, 'An Introduction to Textures in Metals', 1979, London, The Institute of Metals.

27.   J.C. Blade, J. Aust. Inst.Met., 1967, 12, 55.

28.   P.A. Beck, Trans. AIME, 1951, 191, 475.

THE INFLUENCE OF PARTICLES AND DEFORMATION STRUCTURE

ON RECRYSTALLIZATION

D.J. Lloyd

Alcan International Limited
Kingston Laboratories
P.O. Box 8400
Kingston, Ontario
K7L 4Z4

Summary

The early stages of recrystallization have been examined in a high purity Al-Mg alloy and in the commercial AA-5083 alloy. In the essentially particle free Al-Mg alloy recrystallization nuclei are developed from sub-grains which grow in regions of high deformation. In the commercial AA-5083 alloy recrystallization is mainly nucleated at constituent particles which have an associated deformation zone due to plastic incompatibility between the particle and matrix. Particle sizes of about 1 μm are required for nucleation. While both alloys generate shear bands during cold rolling, these are not a major source of recrystallization nuclei.

Introduction

The influence of particles on the recrystallization process has been considered by a number of workers (1). Because coarse particles cannot undergo the equivalent plastic strain of the matrix, accommodation strains are generated in the matrix around them and recrystallization is nucleated in these regions (2). On the other hand fine particles can exert a pinning effect through Zener drag and inhibit recrystallization (3).

Many commercial alloys contain both coarse constituent particles, which are produced during solidification, and finer dispersoid particles which can exert a drag. As a result both particle effects can potentially occur. In addition, in the high solute containing alloys, strain becomes localized into shear bands at high rolling reductions (4). This means that additional strain gradients are present in the matrix which are not necessarily associated with particles and are potential sites for the nucleation of recrystallized grains.

In this paper the recrystallization of a relatively high purity Al – 4.5 Mg alloy, which contains only a low concentration of constituent particles (Al-Fe-Si phase) but not dispersoids, and of commercial AA-5083 alloy is considered. This alloy contains coarse constituents and finer α Al-Fe-Si and $MnAl_6$ dispersoid particles.

Experiment

The compositions of the alloy are shown in Table I.

Table I. Element in Wt.%

Alloy	Mg	Fe	Si	Mn	Cr
Al-Mg	4.5	0.1	0.1	<0.001	<0.001
AA-5083	4.4	0.4	0.4	0.7	0.05

Fully annealed material was cold rolled 70% and then annealed at 260 and 280°C. The recrystallization was followed optically and by electron microscopy. For the optical microscopy the heat treated sample was first decorated with Mg_2Al_3 precipitate by holding at 125°C for 7 days and then etched in 10% H_3PO_4 for 10 to 15 minutes. The specimen plane perpendicular to the rolling plane and containing the rolling direction has been examined in both cases.

Results

Al-Mg Alloy

The Al-Mg cold rolled alloy shows extensive shear banding expected in high solute containing alloys, Figure 1. In transmission electron microscopy the deformation structure consists of bands of high dislocation density superimposed on a background of relatively uniform dislocation density, Figure 2. The dislocation bands are usually along the traces of the (111) slip planes.

Figure 1 – Shear bands in cold rolled Al–Mg alloy.

Figure 2 – Dislocation banding in cold rolled Al–Mg alloy.

47

After 5 mins at 260°C no major changes in microstructure are apparent optically but after 15 mins at the same temperature many small grains have formed, Figure 3.

(a)

(b)

Figure 3 – Microstructure in cold rolled Al–Mg after 15 minutes at 260°C

Some of these grains are aligned along the direction of the shear bands but this is not a predominant mechanism and most of the shear bands have disappeared. The majority of new grains are associated with original grain boundaries, many of which are becoming serrated due to boundary migration. After 30 mins at 260°C the recrystallized grains are larger and from the shape of these grains it is clear that extensive grain boundary migration is occurring, Figure 4. With increasing time recrystallization progresses and is complete after 180 mins.

Figure 4 - Microstructure in cold rolled Al-Mg
after 30 minutes at 260°C.

At higher temperatures similar processes occur but the kinetics are faster. An interesting aspect occurred under all conditions but was particularly apparent at higher temperatures. As shown in Figure 5 there was the occasional highly elongated grain which was more resistant to recrystallization than the average. It appears that nucleation is difficult in these grains and only a limited number of grain orientations are able to migrate into them, Figure 6. It is possible that the elongated grains were developed from an initial orientation which quickly established a stable deformation orientation and as a result does not have as high a stored energy as the majority of the grains. Eventually, however, these grains do recrystallize because there is no evidence of them in the later stages of recrystallization. This can be seen from Figure 7 which is the grain structure after 60 min at 280°C.

Transmission electron microscopy shows that extensive recovery has occurred after 5 mins at 260°C, which was not apparent from optical microscopy. A very fine sub-grain structure has developed with the occasional very clean, sharp, sub-grain, Figure 8. Convergent beam microdiffraction (CBM) from these sub-grains and the surrounding regions showed that they had the same orientation but were rotated. In some cases the sharp sub-grains were aligned at an angle to the rolling direction, consistent with shear band nucleation, Figure 9. After 15 mins at 260°C many more sharp sub-grains have developed and large grains with high angle boundaries are also present in the microstructure, Figure 10. After 30 mins many large, high angle grain

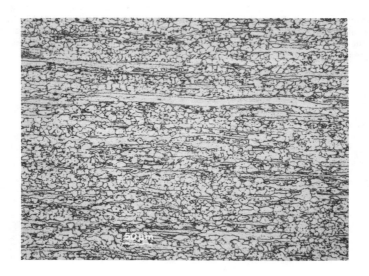

Figure 5 – Elongated grains resistant to recrystallization.

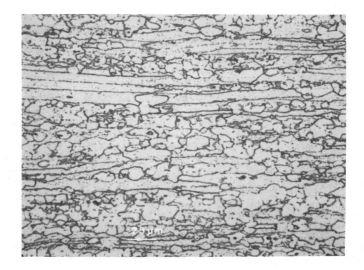

Figure 6 – Grain boundary migration into elongated grains.

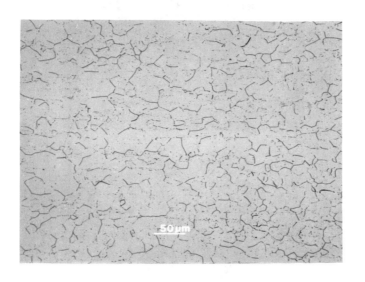

Figure 7 – Microstructure after 60 minutes at 280°C.

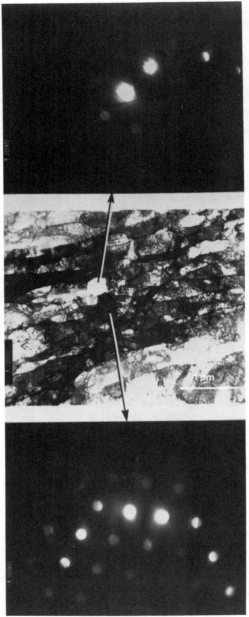

Figure 8 – Sub-grain structure after 5 minutes
at 260°C.

Figure 9 – Sub-grains aligned along direction of dislocation bands.

Figure 10 – Both sub-grains and high angle grain boundaries present after 15 minutes at 260°C.

boundaries have developed, often occurring in rows or clusters, Figure 11. In some cases the grain boundaries are quite sharp, consistent with growth by the low angle sub-boundary absorption. In other cases the grain boundaries show cusps, indicative of general migration, Figure 12.

Figure 11 – Grain structure after 30 minutes at 260°C.

Figure 12 – Example of general grain boundary migration.

As was noted previously this alloy contains some constituent, which occur as short stringers. In these regions recrystallization occurs as a row of grains along the length of the stringer with the grains elongated in the direction of particle alignment. Diffraction showed that many of the adjacent grains had the same orientation but with a relative rotation, Figure 13.

AA-5083

Cold rolling alloy AA-5083 also develops shear band, some of which are macroscopic in that they extend across a large number of grains, Figure 14. One feature of this alloy which is different from the Al-Mg is the presence of about 2 volume percent of constituent particles, aligned in the rolling direction. These particles are associated with a small intense deformation zone resulting from the plastic strain incompatability between the particle and the matrix, Figure 15. In the transmission electron microscope the dislocation structure is heavily banded, Figure 16, and the deformation structure is heavily distorted in the region of constituent particles. Figure 17 shows the rotations that occurred around one of the constituent particles.

After 15 min at 260°C the shear banding has disappeared and recrystallized grains have extensively nucleated throughout the material, Figure 18. The majority of recrystallized grains are associated with particles and the coarsest particles often nucleate several grains. As the annealing time increases additional grains are nucleated together with limited grain growth, Figure 19. Recrystallization is essentially complete after 60 min at 260°C, Figure 20.

Transmission electron microscopy confirms the optical microscopy with recrystallization nucleating at particles and coarser particles often being associated with several recrystallized grains, Figure 21. Most of the small grains observed could be associated with a nucleating particle. Migrating boundaries between recrystallized grains often show cusps and their motion is clearly being inhibited by the dispersoids, Figure 22.

Discussion

In the Al-Mg recrystallization nuclei are being developed from sub-grains generated in regions of high strain. These regions can be shear bands but more often the high strain regions are in the neighbourhood of original grain boundaries. There are also many examples of original grain boundaries bowing out and migrating into high deformation regions. Clearly large strain gradients are being developed adjacent to the grain boundaries. Many of the shear bands extend across grain boundaries and such intersection points would be expected to be high defect regions.

In the case of AA-5083 coarse particles are the major nuclei. In Figure 23 the frequency of nucleation with particle size is plotted. No nucleation is observed for particles below 0.5 µm and the majority of events is associated with particles greater than 1 µm. The frequency decreases above 2 µm because the fraction of particles present in the alloy decreases at larger sizes.

The growth of nucleated grains and of sub-grains with annealing time at 260°C for Al-Mg is shown in Figure 24. (These were obtained from TEM.) Extensive growth of grains occurs initially but then slows down after about 30 min. This reduction in kinetics would be expected as the stored energy decreases with time. A similar plot is shown for AA-5083 in Figure 25, and compared with Figure 24 the kinetics are slower. The temperature considered for AA-5083 is 20°C below that for Al-Mg but these are the temperatures where recrystallization is completed in equivalent

Figure 13 – Row of recrystallized grains nucleated
along a stringer of constituent particles.

Figure 14 – Macroscopic shear bands in cold rolled
AA-5083 alloy.

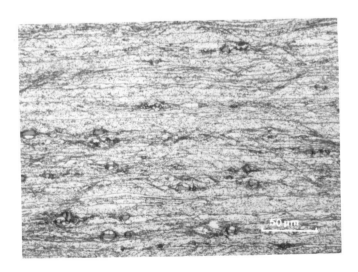

Figure 15 – Deformation zone associated with constituent
particle in AA-5083.

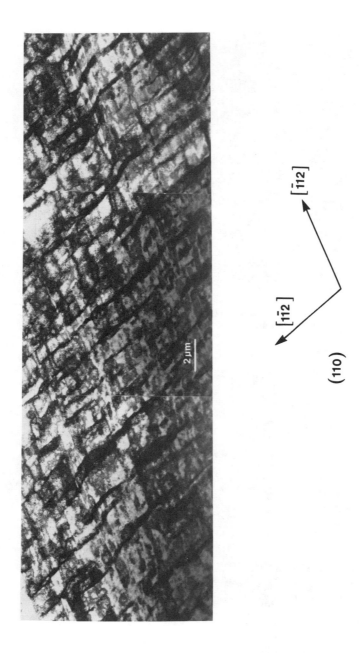

Figure 16 – Banded dislocation structure in cold rolled AA–5083.

Figure 17 – Lattice rotations associated with a
constituent particle. (Arrows indicate [202] direction.)

Figure 18 – Microstructure of cold rolled AA–5083 after 15 minutes at 260°C.

Figure 19 – Microstructure of cold rolled AA–5083 after 30 minutes at 260°C.

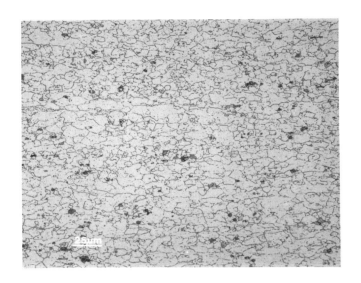

Figure 20 – Grain structure after 60 minutes at 260°C.

Figure 21 – Recrystallization nucleated at particles.

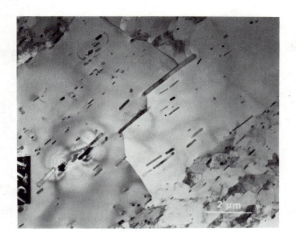

Figure 22 – Dispersoids inhibiting grain boundary motion.

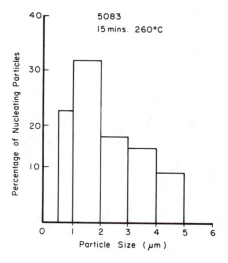

Figure 23 – The frequency of nucleation against particle size in AA-5083.

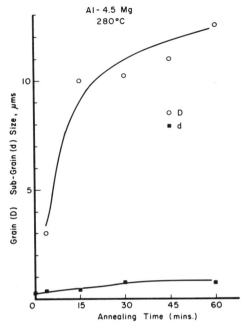

Figure 24 - The growth of nucleated grains in the Al-Mg alloy.

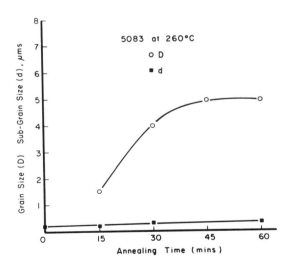

Figure 25 - The growth of nucleated grains for AA-5083 alloy.

times. The lower kinetics and smaller grain sizes is consistent with the dispersoids exerting a drag on migrating boundaries. The sub-grain growth is also more rapid in Al-Mg than in AA-5083 which exhibits very low growth.

The driving force for recrystallization is the stored energy of cold work, or more specifically, the gradients in stored energy. This is not very well known but from stored energy measurements $F_R = 10^6 - 10^7$ N/m^2 is reasonable. This driving force has to overcome particle pinning due to Zener drag F_p, and grain boundary energy effects F_G. For nucleation growth

F.s   *Fdrive +*

$$F_R > F_p + F_G \tag{1}$$

(Stored energy)

= Stain stress energy

— relaxed energy.

$$> \frac{3f\gamma}{2r} + \frac{2\gamma}{R} \tag{2}$$

where f is the volume fraction of pinning particles, γ is the grain boundary energy, approx. 1 J/m^2, r is the pinning particle radius and R is the grain radius. In the case of Al-Mg there is no Zener pinning and $2\gamma/R$ would be of the order of $10^3 - 10^4$ N/m^2, so nucleation growth from small sub-grains is viable. In AA-5083 the dispersoids exert a Zener drag and larger diameter nuclei are required. However the constituent particles provide nuclei of 1 to 2 μm diameters and these can overcome the pinning under the stored energy driving force. As pointed out by Chan and Humphreys (5), dispersoids on sub-grain boundaries exert an enhanced drag and in those regions where recrystallization has occurred, high angle boundaries are migrating only under the driving force of F_G. These two effects in combination explain why grain growth kinetics are reduced in AA-5083 relative to Al-Mg.

A final point of interest is the orientation of recrystallized nuclei. In the plane of examination the predominant orientation of the deformed matrix is (110) in both alloys. Figure 26 shows that this is also a major orientation of new grains in Al-Mg and Figure 27 is a similar plot in AA-5083. Therefore, even though the nuclei are associated with coarse particles in AA-5083 the distribution of new grain orientations are very similar in both alloys. It is also apparent that the nucleated grains tend to inherit the deformed orientation. The final texture developed will depend on subsequent growth kinetics. The growth of individual grains will depend on their local environment, including stored energy effects, orientation of surrounding grains and grain shape considerations. These factors are outside the scope of the present paper.

Conclusions

In Al-Mg alloy with low particle contents the recrystallization nuclei are developed from sub-grains which grow in high deformation regions. These regions can be associated with shear bands but more often are in original grain boundary regions. In the AA-5083 coarse particles with their associated deformation zones are the main regions of nucleation. Particles with diameters of 1 μm and larger are effective as nuclei. Finer dispersoids inhibit the growth of new grains. Again shear bands are not a major source of nuclei. The crystallographic orientation of the nuclei is often inherited from the orientation developed during cold working.

Acknowledgement

The author is grateful to Alcan International Limited for permission to publish

64

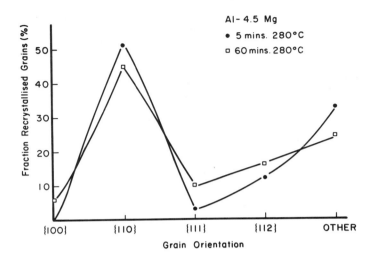

Figure 26 – Orientation distribution of recrystallized grains in Al-Mg alloy.

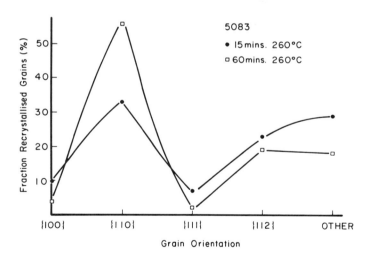

Figure 27 – Orientation distribution of recrystallized grains in AA-5083 alloy.

this paper; and to his colleagues in the Kingston Research Centre for their help in many aspects of the work.

References

(1)   "Recrystallization and Grain Growth of Multi-Phase and Particle Containing Materials" (ed. N. Hansen, A.R. Jones and T. Leffers), Proc. 1st Risö International Symposium on Metallurgy and Materials Science, 1980, Risö National Laboratory, Roskilde, Denmark.

(2)   F.J. Humphreys: Acta Metall., 25 (1977) p. 1323.

(3)   C. Zener: quoted by C.S. Smith: Trans TMS-AIME, 175 (1949) p. 15.

(4)   D.J. Lloyd, E.F. Butryn and M. Ryvola: Microstructural Science, 10 (1982) p. 373.

(5)   H.M. Chan and F.J. Humphreys: Acta Metall., 32 (1984) p. 235.

THERMOMECHANICAL PROCESSING OF HEAT-TREATABLE

ALUMINUM ALLOYS FOR GRAIN SIZE CONTROL

John A. Wert*

Department of Materials Science
University of Virginia
Charlottesville, VA 22901

Summary

    Reduction of grain size of aluminum alloys can increase strength and can lead to superplastic behavior during elevated temperature deformation. A brief review of the microstructural requirements for grain boundary strengthening and for superplasticity shows that special thermomechanical processing (TMP) methods are required to produce the fine grain sizes needed to obtain these properties. The key to TMPs designed for grain size control is managing recrystallization with the aid of particle dispersions. Recrystallization of heat-treatable aluminum alloys can proceed via two paths. Controlling grain size through discontinuous recrystallization requires a bimodal particle size distribution to provide nucleation sites for recrystallizing grains and to restrict grain growth after recrystallization. Continuous recrystallization requires a particle dispersion to provide a large Zener drag pressure, suppressing the more rapid discontinuous process. Our understanding of discontinuous recrystallization is sufficient to permit a model to be developed for prediction of recrystallized grain size as a function of alloy and processing parameters. This model is described in the paper, and it is applied to the problem of optimizing alloy composition and processing conditions to produce fine grain sizes.

*Formerly: Rockwell International Science Center, Thousand Oaks, CA 91360

67

1. Introduction

For the past 50 years, strength improvements in high strength aluminum alloys have been sought through increasingly careful control of precipitate dispersions. A number of alloy classes have evolved from these efforts, the most prominent being alloys based on Al-Cu, Al-Mg-Si, Al-Zn-Mg and Al-Li. However, recent developments indicate that grain refinement is an alternate method for increasing strength. Moreover, superplastic behavior is found during elevated temperature deformation of aluminum, provided that a fine grain size can be stabilized at the deformation temperature.

Processing of heat-treatable aluminum alloys to produce fine, stable grain structures has required advances in understanding of basic aspects of recrystallization in the presence of particle dispersions. This paper reviews recrystallization in the presence of particle dispersions, and considers how recrystallized grain size may be controlled through selection of suitable alloy compositions and thermomechanical processing (TMP) methods.

2. Microstructural Requirements

In Part 2 of this paper, microstructural requirements for strengthening and superplastic formability are examined. These requirements lead to consideration of processing methods that can provide the necessary microstructural characteristics.

2.1 Grain Refinement for Strengthening

Experimental and theoretical studies have clarified the relationship between grain size and flow stress in aluminum alloys [1-6]. Most of the experimental data can be described by the Hall-Petch relationship,

$$\sigma = \sigma_0 + kd^{-\frac{1}{2}} \qquad (1)$$

where σ is the yield stress, σ_0 is a constant, d is the average grain diameter, and k is the Hall-Petch constant. Values of k near 0.1 $MPa \cdot m^{\frac{1}{2}}$ have been found for pure aluminum and for a variety of aluminum alloys, as shown in Table I. To achieve a 100 MPa contribution to flow stress from grain boundary strengthening, a change in d from 100 μm to 0.9 μm would be necessary.

The effect of subgrain or grain boundary misorientation on grain boundary strengthening has been discussed in recent reviews [7-9]. Some experimental evidence suggests that subgrain boundaries contribute less to strength than grain boundaries, while other evidence supports the idea that strengthening is independent of boundary misorientation. Thompson [9] concluded that strengthening is approximately independent of grain boundary misorientation when the grain size exponent in the Hall-Petch relationship is near -0.5, as is usually observed when grains or well-recovered subgrains are present.

2.2 Grain Refinement for Superplasticity

Deformation of aluminum alloys with fine grain size at elevated temperatures and modest strain rates leads to superplastic behavior. It is now generally agreed that superplasticity in metals occurs when grain

TABLE I

Summary of Hall-Petch Data for Recrystallized Aluminum Alloys

Alloy	Strain	k, MPa·$m^{\frac{1}{2}}$	Reference
Al-Mg	Yield	0.26	[1]
Al	0.005	0.070	[2]
Al	0.002	0.065	[3]
Al-Ni	Yield	0.14	[4]
	0.05	0.13	
	0.10	0.12	
Al	0.002	0.034	[5]
	0.005	0.034	
	0.001	0.036	
	0.002	0.042	
Al-Zn-Mg-Cu	0.002	0.12	[6]
	0.01	0.082	
	0.02	0.080	
	0.04	0.087	
	0.06	0.087	
	0.08	0.093	

boundary sliding provides a large contribution to the total deformation, giving rise to a strong dependence of superplastic properties on grain size [10-13]. As an example, Fig. 1 shows flow stress as a function of strain rate for 7475 Al processed to several grain sizes prior to straining at 516°C [13]. Maximum superplastic elongation is generally found at a strain rate corresponding to the maximum slope of the log σ vs log ε relationship. For 7475 Al, Fig. 1 shows that grain sizes less than 20 μm are required to obtain superplasticity at strain rates above 10^{-5} s^{-1}. Since higher strain rates translate directly into lower component forming times, finer grain sizes that confer superplasticity at higher strain rates are clearly desirable.

Superplasticity in aluminum alloys requires deformation at temperatures above approximately 0.8 of the absolute melting point, a regime where rapid grain growth normally occurs [12]. However, rapid grain growth quickly changes the optimum strain rate for superplasticity to such a low value that continued deformation at an optimum strain rate for superplasticity would be impractical. Thus, heat-treatable aluminum alloys designed for superplastic forming must include particle dispersions capable of stabilizing the grain structure at elevated temperatures.

A third microstructural prerequisite for superplasticity is presence of high-angle grain boundaries. The effect of adjacent grain misorientation on boundary sliding rate has been demonstrated in beautifully detailed investigations carried out by Bisconti and coworkers [14,15]. For example, their findings show that substantial sliding of [100] tilt boundaries in pure aluminum requires a misorientation of more than 10° between adjacent grains. Thus, microstructures designed for superplastic forming must contain predominantly high-angle grain boundaries. The effect of average boundary misorientation on superplastic properties of an Fe-Ti-Ni alloy has been demonstrated by Benedek and

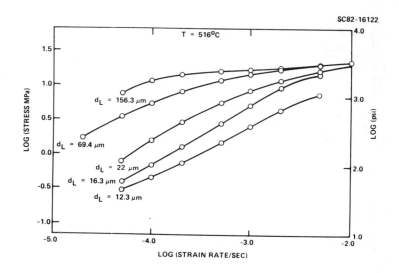

Fig. 1 - Flow stress as a function of strain rate and mean linear
intercept longitudinal grain size (d_L) for 7475 strained at 516°C.
(Courtesy C. H. Hamilton).

Doherty [16].

3. Recrystallization Processes in Aluminum Alloys

Grain refinement of heat-treatable aluminum alloys can be achieved
only through recrystallization, in contrast to steels and titanium alloys
where phase transformations can be used in conjunction with
recrystallization to achieve fine grain sizes [17-19]. To attain grain
sizes smaller than 20 μm in heat-treatable aluminum alloys, advances in
basic understanding of the effects of particles on recrystallization have
been required. From this work, it has become apparent that two distinct
recrystallization paths can lead to recrystallized microstructures with
fine grain sizes [17,19-23]. In Part 3 of this paper, use of particle
dispersions to select a recrystallization path and control recrystallized
grain size is reviewed.

3.1 Particle Effects on Recrystallization

Particle effects on recrystallization are often separated into
consideration of large particles and small particles. However, such a
division is more a reflection of the roles that particles play in the
recrystallization process than of particle size. Particles larger than
about 1 μm in diameter create nucleation sites for recrystallizing grains,
as shown by many previous investigators [24-27]. Creation of potential
nucleation sites for recrystallization adjacent to nondeformable particles
arises from supplementary deformation that occurs in these regions. First,
consider the situation far from large particles where elongated dislocation
cells form at moderate to high strains, see Fig. 2a. Except in
widely-spaced transition bands, adjacent dislocation cells have similar
lattice orientations. During annealing, these low-angle dislocation cell

boundaries recover to form low-angle subgrain boundaries, which have low migration rates and are unsuitable for nucleating recrystallized grains [27,28].

However, the situation is different adjacent to large, nondeformable particles where the matrix must undergo additional deformation to accommodate the presence of the particle. Deformation zones form around such particles, as shown in Fig. 2b. Deformation zones contain small dislocation cells that frequently have quite high misorientations with neighboring cells [24]. During annealing, high-angle boundaries form around highly misoriented cells near large particles, and these small cells become nuclei for recrystallizing grains [24,27]. The high-angle boundaries migrate rapidly outward from their initial location near the particle, first consuming the deformation zone of the adjacent particle, then sweeping through the recovering matrix [24-27]. Fig. 3 shows these events occurring in 7075 Al.

As described above, large nondeformable particles create preformed nucleation sites for recrystallization. However, ability of a particle to form a suitable nucleation site depends strongly on particle size, since particle size determines the extent of lattice rotation in the deformation zone. Previous workers have found that particles must be larger than a critical size to provide the lattice rotations necessary to create potential nucleation sites [74-27]. The critical particle size (n_c) depends on degree of deformation and on alloy content, among other factors. For heavy deformation of heat-treatable aluminum alloys, n_c is near 1 μm. This has led to a rough definition for "large" particles – those capable of creating nucleation sites for recrystallization – as particles with diameters greater than 1 μm.

While a minimum particle size is needed to create nucleation sites for recrystallizing grains, particles of all sizes impede migration of boundaries during and after recrystallization. The ability of particles to exert drag pressure on a migrating boundary was recognized by Zener over four decades ago, as described by Smith in a celebrated private communication [29]. Zener derived a relationship between particle dispersion characteristics and drag pressure (P_Z):

$$P_Z = \beta \, \gamma \, f/r \tag{2}$$

where γ is boundary energy, f is particle volume fraction, r is particle radius and β is a constant equal to 3/4. Later investigators have derived alternate expressions for P_Z, however; all contain the same dependence on f/r as Zener's original expression (see review by Nes, Ryum and Hunderi [30]).

Eq. (2) reveals why substantial drag pressures are usually associated with particles much smaller than 1 μm in diameter. Drag pressures above 20 kPa are needed to stabilize grain sizes of 10 μm. In heat-treatable aluminum alloys, the volume fraction of particles that are insoluble at normal solution treatment temperatures is rarely more than 0.01. Using a value of 0.3 J/m^2 for γ [31], Eq. (2) shows that average particle radius for particles providing boundary drag pressure must be smaller than 0.1 μm. When larger particles are used to provide boundary drag pressure, as in alloys of eutectic composition, particle volume fractions must exceed 0.1 [12,18,19]. In this case, properties of the alloys are substantially dictated by the particle dispersion, and high strength levels sought in heat-treatable aluminum alloys are usually not attained.

71

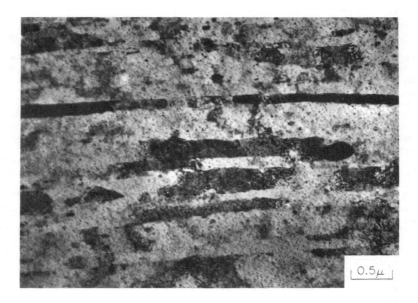

0.5μ

2a

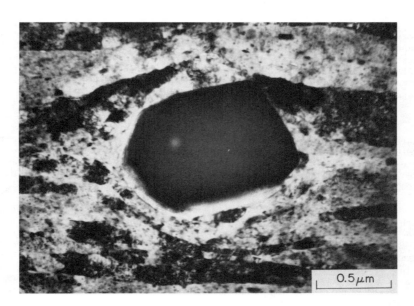

0.5μm

2b

Fig. 2 - Dislocation cell morphology in 7075 Al rolled to a strain of -2.45 at 225°. a) Matrix.   b) Near M-phase precipitate.

72

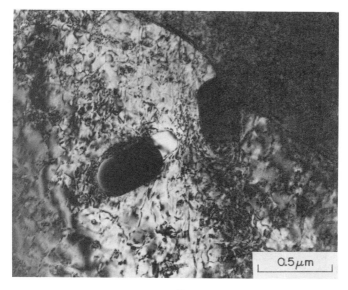

3a

3b

Fig. 3 - Early stages of recrystallization in 7075 Al.

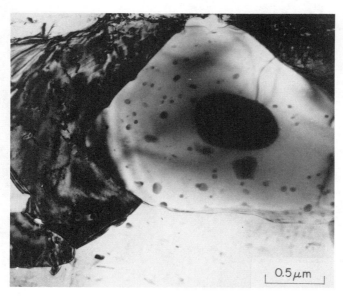

3c

Fig. 3 (contd.) - Early stages of recrystallization in 7075 Al.

Particles which exert substantial drag pressure on migrating
boundaries have several effects on recrystallization processes. They
interfere with nucleation by preventing migration of the high-angle
boundary that forms around potential nuclei. If the drag pressure is
sufficiently high, nucleation of recrystallizing grains is completely
suppressed. When nucleation of recrystallizing grains does occur,
expansion of the new grains into the deformed matrix is slowed by drag
pressure on the migrating recrystallization front. Finally, after
recrystallization is complete, grain growth can be retarded by a dispersion
of insoluble particles in the matrix.

Following the preceding review of the effects of particles on general
aspects of recrystallization, we shall consider the types of particle
dispersions required to control recrystallized grain size for each of the
recrystallization paths.

3.2 Grain Refinement by Discontinuous Recrystallization

This is the classical recrystallization path which proceeds by
nucleation of new grains in the deformation substructure and growth of
these grains until impingement. Boundaries (recrystallization fronts) that
sweep through the deformed microstructure are generally high-angle
boundaries, which have much higher migration rates than low-angle
boundaries [28,32]. This has led previous workers to define
recrystallization as the annealing stage during which high-angle boundary
migration occurs, in contrast to recovery where high-angle boundary
migration does not occur [33]. Since passage of a high-angle boundary
leads to an abrupt, or discontinuous, change in defect density and lattice

74

orientation, the term discontinuous recrystallization has been used to describe this process. For more extensive consideration of mechanistic and kinetic aspects of discontinuous recrystallization, readers are referred to recent reviews of these topics [33,34].

It should be apparent that controlling grain size by discontinuous recrystallization simply requires:

    i.   Providing a sufficient density of nucleation sites for recrystallization,

    ii. Ensuring that many of the potential nucleation sites are activated during recrystallization,

  iii. Providing a dispersion of insoluble particles to stabilize the grain structure after recrystallization.

From the discussion in Section 3.1, condition i requires introducing a dispersion of micrometer-size particles prior to deformation. If a recrystallized grain size of 10 μm is sought, the average particle spacing should be somewhat less than 10 μm to account for potential nucleation sites that are not activated. Such a particle dispersion is conveniently provided by precipitating the soluble alloying elements (Zn, Mg and Cu and in the case of 7XXX alloys) as coarse, equilibrium-phase particles.

Conditions ii and iii are counterposed: ii requires that high-angle boundary motion not be restricted by particle drag pressures, while iii requires a dispersion of grain-size stabilizing particles. As we shall see in Part 4 of this paper, an intermediate value of particle drag pressure can be selected to accommodate both requirements. This approach is possible because the driving pressure for recrystallization is much greater than that for grain growth. Thus, a particle drag pressure sufficient to completely suppress grain growth is small compared to the driving pressure for recrystallization and does not severely restrict nucleation of recrystallizing grains.

In conventional heat-treatable aluminum alloys, dispersoid particles formed by Cr, Zr or Mn additions can provide the boundary drag pressure required for condition iii. In advanced alloys, such as P/M materials, many other possibilities exist for creating suitable particle dispersions to provide boundary drag pressure for grain size stabilization.

3.3 Grain Refinement by Continuous Recrystallization

Continuous recrystallization is an alternate recrystallization path which does not proceed by nucleation and growth of new grains. Instead, continuous recrystallization occurs by gradual subgrain growth, leading to formation of high-angle grain boundaries without requiring high-angle boundary migration [20-23]. The mechanism which allows subgrain growth or coalescence to produce high-angle boundaries has not been well established, although a process similar to that shown in Fig. 4 can be hypothesized. The starting microstructure, shown in Fig. 4a, is composed of elongated dislocation cells introduced during deformation, with adjacent dislocation cells having small misorientations. Prolonged annealing permits subgrain growth by subgrain boundary migration, allowing several low-angle subgrain boundaries to combine into one higher-angle grain boundary, shown in Figs. 4b and 4c. Repetition of this process throughout the microstructure ultimately produces grains separated by moderately high-angle boundaries, although no high-angle boundary motion has occurred.

Although an orderly progression of dislocation cell orientations shown in Fig. 4a is highly idealized, experimental evidence for this process has been obtained by Nes [35]. Nes has measured subgrain size and adjacent subgrain misorientation in an Al-6%Cu-0.3%Zr alloy as a function of time during simultaneous straining and annealing at temperatures near 450°C. Adjacent subgrain misorientation increased from an average of 5° at 10% strain to 14° at 50% strain, with a concurrent increase in subgrain size. If annealing is carried out without simultaneous straining, Nes found that changes in misorientation and subgrain size occur much more slowly [35].

To obtain continuous recrystallization, the comparatively rapid discontinuous recrystallization process must be suppressed. This requires boundary drag pressures sufficient to prevent nucleation of discontinuous recrystallization. Particle dispersions with P_z in excess of 40 kPa can provide the necessary drag forces, suggesting that continuous recrystallization should only be observed in alloys containing very high densities of small particles. Indeed, the most well known examples of continuous recrystallization in aluminum are alloys containing high densities of Al_3Zr dispersoids [36-41] or PM alloys containing high densities of several types of fine particles.

With such high boundary drag pressures needed to suppress discontinuous recrystallization, it seems surprising that sufficient migration of low-angle boundaries to produce continuous recrystallization can occur. However, since boundary drag pressure depends on boundary tension (γ), the drag pressure on a low-angle boundary is less than on a high-angle boundary for a given value of f/r. A recent experimental study by Tweed, Ralph and Hansen [42] confirmed that drag pressure on low-angle boundaries in aluminum is lower than that on high-angle boundaries, although the difference is not as large as predicted by the Zener approach.

Other mechanisms which could allow low-angle boundary migration in the presence of high drag pressure have been proposed. Alborn, Hornbogen and Koster [20,23] suggested that subgrain boundary migration can occur when

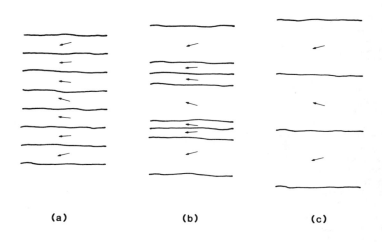

(a)   (b)   (c)

Fig. 4 – Schematic diagram showing accumulation of
misorientation during subgrain growth.

particle coarsening leaves some subgrain boundaries weakly restrained, as illustrated in Figs. 5a and 5b. Alternatively, straining during annealing may provide added driving pressure for boundary migration, accelerating continuous recrystallization. Work by Nes cited previously demonstrates the effect that elevated temperature straining can have on the kinetics of continuous recrystallization [35].

The paucity of observations of continuous recrystallization probably stems from the requirement of achieving an extremely delicate balance: discontinuous recrystallization must be restrained, yet sufficient low-angle boundary migration allowed to obtain the necessary microstructural evolution. Recrystallized grain sizes that are obtained by continuous recrystallization may depend on details of the initial deformation substructure as well as on characteristics of boundary migration. More experimental work is needed to understand the mechanisms of continuous recrystallization and to identify factors that control continuously recrystallized grain size.

3.4 Selection of a Recrystallization Path for Grain Refinement

The previous sections have presented two recrystallization paths that are available for grain refinement in aluminum alloys. When designing a TMP or an alloy/TMP combination, a choice of recrystallization path must be made at the outset because the microstructural requirements for discontinuous recrystallization differ from those for continuous recrystallization.

TMPs that use discontinuous recrystallization for grain size control (DR-TMP) are applicable to a wide variety of existing alloys that have dispersoid distributions giving P_z near 10 kPa. Discontinuous recrystallization is rapid and does not require concurrent deformation at the recrystallization temperature. Recrystallized grain size tends to be relatively insensitive to minor variations in alloy composition and process parameters, with the exception of heating rate to the recrystallization temperature [43]. Material destined for superplastic forming applications can be recrystallized and evaluated before components are formed. However, to offset these benefits, known DR-TMFs cannot produce grain sizes as fine as those that can be achieved through a properly controlled continuous recrystallization process.

(a)   (b)

Fig. 5   Continuous recrystallization by migration of boundaries released as a result of particle coarsening. (After Alborn, Hornbogen and Koster, J. Materials Science, Chapman and Hall, 1969.)

In contrast to DR-TMPs, thermomechanical processes that use continuous recrystallization for grain size control (CR-TMP) generally require special alloy compositions that have dispersoid distributions giving P_z above about 40 kPa. Continuous recrystallization requires either prolonged annealing or concurrent deformation at elevated temperatures, and suppression of discontinuous recrystallization can be sensitive to variations in alloy compositions and process parameters. Material destined for superplastic forming applications is generally supplied in the as-rolled condition and the superplastic properties depend strongly on the initial stages of deformation for each component. Nevertheless, CR-TMPs can produce remarkably fine grain sizes, providing distinct advantages for both strengthening and superplastic forming.

4. Modeling of Discontinuous Recrystallization in Aluminum Alloys

Alloys to be grain refined by discontinuous recrystallization require at least two types of particles: particles to create nucleation sites for recrystallizing grains and particles to restrict grain growth after recrystallization. In heat-treatable aluminum alloys, it is convenient to form the large particle distribution by precipitating solutes such as Cu, Zn or Mg and to use dispersoid particles to limit grain growth. In addition, constituent phase particles are invariably present, and we shall also examine their role in grain size control.

To make use of these particle dispersions, DR-TMP methods designed for grain refinement of heat-treatable aluminum alloys employ the following steps [43-49].

1. Heat treatment to provide appropriate particle dispersions.
2. Deformation below the recrystallization temperature.
3. Annealing to permit recrystallization.

External variables in such processes include precipitate particle size distribution (controlled by alloy composition and heat treatment), constituent and dispersoid particle size distributions (controlled mainly by alloy composition), deformation conditions and recrystallization conditions.

Since the number of variables in such a DR-TMP is enormous, models which can predict the effects of these variables on recrystallized grain size can aid TMP development. Such a model should include both recrystallization and grain growth effects so that the stable recrystallized grain size can be predicted for given alloy composition and thermomechnical processing conditions. A model capable of prediciting recrystallized grain size (ignoring grain growth) has been previously published by Nes and Wert [50]. In the remainder of Part 4, the original model is briefly described and is extended to include the effects of grain growth.

4.1 Modeling of Recrystallization-Limited Grain Size

Nucleation of discontinuous recrystallization in an alloy containing particles larger than approximately 1μm in diameter occurs in the deformation zones of the particles. Since the density of such particles usually considerably exceeds the density of alternate nucleation sites at grain boundaries or transition bands, nucleation is assumed to occur only at particles. To act as a potential nucleation site, a particle must exceed a critical diameter, η_c, which is given by [24-27]:

$$\eta_c = \frac{2\gamma}{3(P_D - P_Z)} \qquad (3)$$

where γ is the grain boundary energy, P_D is the driving pressure for recrystallization provided by the stored energy of deformation, and P_Z is the Zener drag pressure from small particles.

Once η_c has been determined, the density of potential nucleation sites is simply the density of particles larger than the critical diameter, $N_v(\eta_c)$. This can be obtained from the size distribution of large particles, which is conveniently found using an approach described by Sandstrom [51]. From Sandstrom's analysis, the density of particles larger than diameter η is given by:

$$N_v(\eta) = \frac{c}{B}\left(\left[\frac{2B\eta}{\pi}\right]^{\frac{1}{2}} \exp(-B\eta) + \frac{3}{4} \; \mathrm{erfc} \; (\sqrt{B\eta})\right) \qquad (4)$$

where B and C are experimental constants which can easily be determined from particle size distribution on a plane.

Calculation of recrystallized grain size from the number of potential nucleation sites makes use of a theory for phase transformations originally proposed by Avrami [52-54]. The Avrami theory applies to materials in which transformation starts at a number of randomly distributed, pre-existing nucleation sites, $N_v(\eta_c)$ in the present case. Each site is assumed to have a nucleation frequency ν. After nucleation, grains are assumed to grow outward into the unrecrystallized matrix at a constant growth rate, G, until impingement.

Recrystallized grain size may be calculated using either of two extreme cases of the Avrami theory: site-saturated transformation kinetics or Johnson-Mehl (J-M) transformation kinetics [55]. Previous observations indicate that J-M kinetics realistically describe recrystallization in Al-Mn alloys [56]. Assuming that J-M kinetics also apply to recrystallization of heat-treatable aluminum alloys, the recrystallized grain size (D_R) is given by:

$$D_R = \left[\frac{G}{\nu N_v(\eta_c)}\right]^{\frac{1}{4}} \qquad (5)$$

Eq. 5 gives the recrystallized grain size in terms of nucelation and growth rates of recrystallizing grains. The growth rate can be re-expressed in terms of a driving pressure and a boundary mobility:

$$G = M \; (P_D - P_Z) \qquad (6)$$

where M is boundary mobility and the term $(P_D - P_Z)$ describes the total driving pressure for boundary motion during recrystallization. Thus, Eq. (5) can be re-expressed as

$$D_R = K \left[\frac{P_D - P_Z}{N_v(\eta_c)}\right]^{\frac{1}{4}} \qquad (7)$$

where $K = (M/\nu)^{\frac{1}{4}}$ is constant for given recrystallization conditions.

From Eq. (7), it appears that four parameters must be established to predict recrystallized grain size with this model.

  i.   Size distribution of large particles.
  ii. Stored energy of deformation.
  iii. Zener drag provided by dispersoid particles.
  iv. The constant K.

Previous papers have discussed validity of the various assumptions implicit in Eq. (7) and application of this model to thermomechanical processing of 7075 Al [50,57,58]. The model was able to accurately predict the variation of recrystallized grain size with amount of deformation for 7075 Al with several large particle size distributions.

4.2 Modeling of Grain Growth Limited Grain Size

After recrystallization is complete, grain growth may occur at the recrystallization temperature. Recrystallization of heat-treatable aluminum alloys is usually carried out at or near the solution treatment temperature where soluble alloying elements are in solid solution. Thus, only insoluble constituent and dispersoid particles are available to limit grain growth. Constituent particles are too large and widely spaced to provide significant grain size stability, so we must rely on dispersoid particles to achieve the needed stability.

Assuming that abnormal grain growth does not occur, the limiting grain size during grain growth is given by an expression originally derived by Zener [29]:

$$R_G = \frac{1}{\beta} \left(\frac{r_D}{f_D} \right) \tag{8}$$

where f_D and r_D are dispersoid volume fraction and average radius, R_G is the limiting grain radius and β is a previously-defined constant. Zener's derivation provided a value of 3/4 for β. Available experimental evidence indicates that the value of 3/4 may be too large by a factor of 2 or 3 (for high-angle grain boundaries) [42,59]. Assuming that drag pressure is 2 times larger than that proposed by Zener, the limiting grain diameter during normal grain growth (D_G) may be written:

$$D_G = \frac{4}{3} \left(\frac{r_D}{f_D} \right) \tag{9}$$

Note that a smaller value for β also alters Eq. 2 which is used to determine the drag pressure due to a particle dispersion:

$$P_Z = \frac{2}{3} \gamma \frac{f_D}{r_D} \tag{10}$$

The value of P_Z is needed to calculate the recrystallization limited grain size, D_R, as discussed in the previous section.

4.3 Grain Size Prediction Using the Model

This section illustrates application of the model to prediction of stable grain sizes for thermomechanical processing of Al-Zn-Mg-Cu-Cr alloys

TABLE II

Constants for Grain Size Prediction, 7075 Al

Constant	Value	Units	Reference
K	0.0056	$(m/Pa)^{1/4}$	[50]
γ	0.30	N/m^2	[50]
f_D/r_D	5.0×10^4	m^{-1}	–
B	5.3×10^6	m^{-1}	–
C	3.8×10^{24}	m^{-4}	–

TABLE III

Variation of P_D with Strain for 7075 Al [50]

Strain	P_D, kPa	η_c, μm	$N_v(\eta_c)$, m^{-3}	D_R, μm
1.0	100	2.2	1.8×10^{13}	50
1.5	140	1.7	2.1×10^{14}	29
2.0	170	1.4	1.0×10^{15}	20
2.5	200	1.1	3.9×10^{15}	15
3.0	230	0.98	7.9×10^{15}	13
3.5	250	0.87	1.4×10^{16}	12
4.0	280	0.77	2.2×10^{16}	11

similar to 7075 Al. Initially, presence of constituent particles formed by Fe and Si impurities will not be included in the analysis. The constants required for application of the model are listed in Tables II and III. In the present paper, calculated and experimental grain sizes represent average grain size in 3 dimensions, not mean linear intercept sizes on a plane.

Results of model can be graphically presented by plotting grain size as a function of strain, as shown in Fig. 6. The solid line represents D_R as a function of strain while the broken line represents D_G, which is independent of strain. The lines cross at a critical value of strain, ϵ_c. For $\epsilon < \epsilon_c$, recrystallized grain size exceeds the grain growth limited grain size. Under these conditions, grain growth does not occur after recrystallization and the stable grain size is recrystallization limited. For $\epsilon > \epsilon_c$, recrystallized grain size is smaller than D_G. In this case grain growth to D_G will occur after recrystallization is complete and final grain size is grain-growth limited. The value of ϵ_c in this example corresponds to a rolling reduction of 94%, a larger reduction than is commonly employed in thermomechanical processing.

The results shown in Fig. 6 are useful for DR-TMP development because they display stable grain size as a function of strain, a processing variable. However, alternate representations of the modeling results are more useful for alloy selection. For example, Fig. 7 shows grain size as a function of f_D/r_D (an alloying variable) for a TMP strain of 2.5. The grain-growth-limited grain size (broken line) decreases with increasing f_D/r_D. However, the recrystallization-limited grain size (solid line) increases with increasing f_D/r_D. This is caused by deactivation of some potential nucleation sites due to restriction of grain nucleation by dispersoid particles. As in Fig. 6, stable grain size is given by the

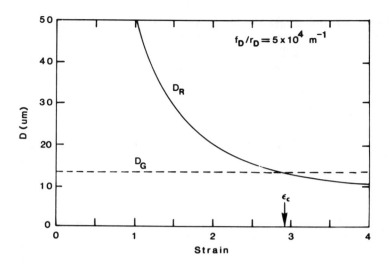

Fig. 6   Recrystallization limited and grain-growth-limited grain size as functions of rolling strain.

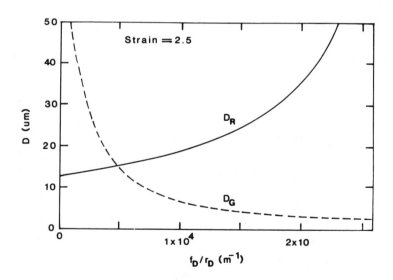

Fig. 7   Recrystallization limited and grain-growth-limited grain size as functions of dispersoid f_D/r_D.

uppermost of the two curves. These results show that an optimum value of f_D/r_D exists for given large particle distribution, deformation conditions and recrystallization conditions.

Obviously, results of this model can be used to examine the effects of each alloy and processing parameter on stable grain size. Before proceeding to consider design of alloys and thermomechanical processes for optimum grain refinement, comparison of modeling results with experimental data for 7075 Al can provide confidence in the accuracy of the model.

To properly represent an alloy such as 7075, constituent phase particles must be included in the large particle distribution. This has been done by providing two sets of B and C parameters: B_E and C_E describe the equilibrium M-phase particle distribution, while B_C and C_C describe the constituent particle dispersion. B_C and C_C can be determined from as-quenched samples while B_E and C_E are determined after heat treatment. These values of B and C are listed in Table IV; other parameters retain the values used previously.

TABLE IV

Particle Size Distributions for 7075 Al

Particle Type	Parameter	Value
M-Phase	B_E C_E	5.3×10^6 m^{-1} 3.8×10^{24} m^{-4}
Constituent	B_C C_C	1.5×10^6 m^{-1} 8.4×10^{21} m^{-4}

Fig. 8 shows modeling results for 7075 Al with both M-phase particles and constituent particles available to create nucleation sites for recrystallizing grains. Recrystallization-limited grain size is represented by the solid line while the broken line shows grain-growth limiting grain size. Data points for 7075 Al are in good agreement with the predicted results.

Three regimes can be identified on Fig. 8:

  i. For strains less than 2, grain size is recrystallization limited; constituent particles provide most of the nucleation sites for recrystallizing grains.

  ii. For strains between 2 and 2.8, grain size is recrystallization-limited; M-phase precipitates provide most of the nucleation sites.

  iii. For strains larger than 2.8, grain size is grain-growth-limited.

The good agreement between predicted and experimental grain sizes and other aspects of recrystallization provides confidence that the model described previously can predict stable grain size for DR-TMPs.

5. Application of Model to Design of Alloys and Thermomechanical Processes for Grain Refinement

The DR-TMP model discussed in Part 4 can be used to select optimum alloy characteristics and thermomechanical processing steps for maximum

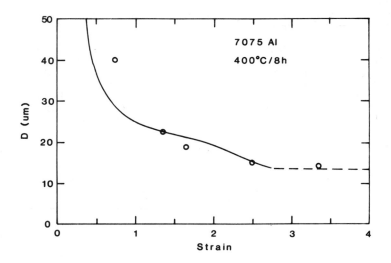

Fig. 8   Theoretical and experimental results for stable grain size in
          7075 Al as a function of rolling strain.

grain refinement. A number of factors enter into the model, and the
influence of each factor on stable grain size can be evaluated. Finally,
these factors can be combined to provide some overall recommendations for
selection of alloy characteristics and DR-TMP steps that are likely to be
beneficial for grain refinement in most situations.

5.1 Equilibrium Precipitate Distribution

In Part 4, all examples of application of the model used a particle
size distribution obtained in 7075 Al by a 400°C/8 h precipitation
treatment. However, broader or narrower particle distributions can be
achieved by alternate treatments. The effect of particle size distribution
on stable grain size is considered in this section.

Fig. 9 shows the four particle size distributions used in this
evaluation. Particle size distribution 1, the baseline distribution,
corresponds to the 400°C/8 h treatment examined previously. Numerical
integration of this distribution shows that the volume fraction of
equilibrium precipitate (f_E) is 0.045. Particle size distributions 2, 3
and 4 were selected to provide identical volume fraction of precipitates,
but broader or narrower size distributions as listed in the upper portion
of Table V.

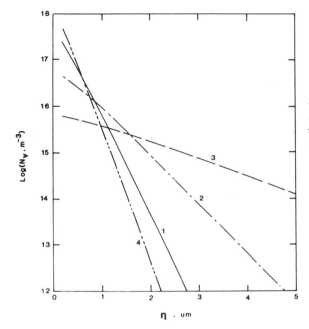

Fig. 9 Particle size
distribution used for
modeling. f_E is 0.045
in all cases.

TABLE V

Particle Size Distribution Parameters

Distribution	B, m^{-1}	C, m^{-4}	f_E
1	5.3×10^6	3.8×10^{24}	0.045
2	2.6×10^6	2.3×10^{23}	0.045
3	1.0×10^6	1.0×10^{22}	0.045
4	7.0×10^6	1.2×10^{25}	0.045
5	5.3×10^6	1.7×10^{24}	0.002
6	5.3×10^6	3.3×10^{24}	0.004
7	5.3×10^6	5.0×10^{24}	0.006
8	5.3×10^6	6.7×10^{24}	0.008

85

Effect of changing particle size distribution on stable grain size as a function of strain can be seen in Fig. 10a. Particle distributions 2 and 3, which are broader than distribution 1, provide better grain refinement at relatively low strains where the critical particle diameter is quite large. In contrast, narrower distributions provide better grain refinement at very high strains where the critical particle size is small. Note that stable grain size is grain-growth limited for distributions 1, 2 and 3 at large strains. Fig. 10b shows the effect of the four particle size distributions on stable grain size as a function of dispersoid f_D/r_D. For f_D/r_D values corresponding to existing alloys (4×10^4 m^{-1} < f_D/r_D < 8×10^4 m^{-1}), particle distribution 2 provides the smallest stable grain size.

These results show that a moderately broad particle size distribution is best for many thermomechanical processing treatments. The 400°C/8 h treatments described as "optimum" by Wert et al [49] do not provide the finest possible grain size in 7075 Al, although the difference between distributions 1 and 2 is quite small at strains above 2.5. In special cases, such as reduction limited to less than 1.5, a very broad particle size distribution could provide considerably finer grain sizes than the narrow distribution created by the 400°C/8 h treatment.

5.2 Equilibrium Precipitate Volume Fraction

Since recrystallization-limited grain size depends on the number of nucleation sites available to nucleate recrystallizing grains, the volume fraction of such precipitates will affect recrystallized grain size. In this section, the effect of volume fraction of equilibrium precipitates (f_E) on stable grain size is examined for fixed particle size distribution width.

The lower portion of Table V lists the particle size distribution parameters used to evaluate the effects of volume fraction. Sandstrom's parameter C can be varied to provide different volume fractions while the parameter B, which controls distribution width, remains constant. Fig. 11 shows the effects of equilibrium particle volume fraction on recrystallized grain size. As anticipated, higher volume fractions are beneficial to grain refinement. However, only a small difference in recrystallized grain size results from increasing f_E from 0.04 to 0.08. This is because recrystallized grain size depends on (nucleation site density)$^{-\frac{1}{2}}$. Thus, doubling precipitate volume fraction only decreases recrystallized grain size by a factor of about 1.2. Furthermore, precipitate volume fractions are limited to less than about 0.10 in most heat-treatable aluminum alloys. Thus, although increases in precipitate volume fraction do benefit grain refinement by DR-TMPs, the effects will generally be modest.

5.3 Constituent Particles

As discussed in Section 4.3, constituent particles can create nucleation sites for recrystallizing grains in the same manner as equilibrium precipitates. However, constituent particles are known to reduce toughness [60,61], to degrade fatigue properties [60], and to markedly reduce superplastic elongation [12]. The volume fraction of constituent particles is normally reduced as much as possible by achieving low iron and silicon levels in high strength alloys.

Using the particle size distribution parameters listed in Table IV, stable grain size has been calculated with and without constituent particles. Fig. 12 shows the result of these calculations. For strains larger than about 2, the model shows that the effect of constituent

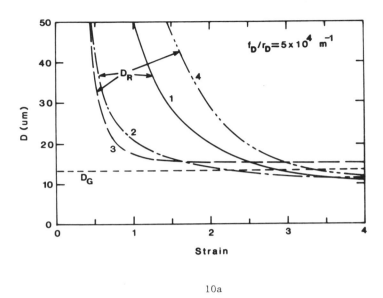

10a

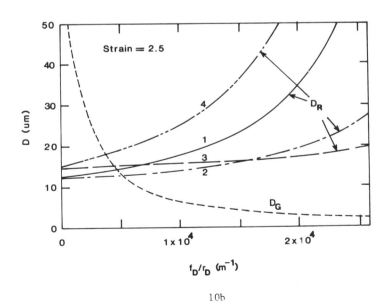

10b

Fig. 10   Effect of particle size distribution on recrystallization-limite
and grain-growth-limited grain size.

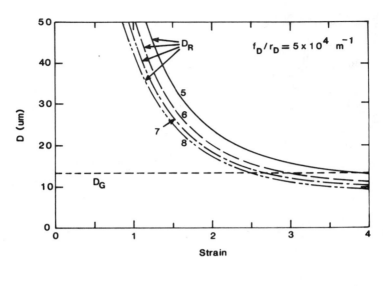

11a

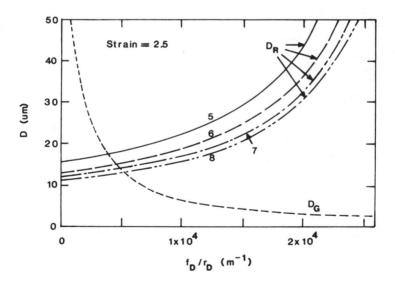

Fig. 11 b

Fig. 11   Effect of equilibrium phase volume fraction (f_E) on
recrystallization-limited and grain-growth-limited grain size.

particles is small. For strains less than 2, constituent particles provide a substantial increase in the number of available nucleation sites. This is because the equilibrium particle size distribution is quite narrow and provides few nucleation sites when the critical particle diameter is larger than about 1.5 μm.

5.4 Optimization of DR-TMP for 7075 Al

Modeling results described in the previous sections suggest several general alloy and DR-TMP characteristics that will promote maximum grain refinement. For alloy selection, a large volume fraction of equilibrium precipitates and dispersoid f_D/r_D values between 6×10^4 and 12×10^4 m^{-1} will benefit grain refinement. Thermomechanical processing treatments should be designed to provide moderately broad large particle dispersions (B near 2 or 3×10^6 m^{-1}), largest possible volume fraction of the equilibrium phase, and deformation strains of 2.5 or larger.

Based on these recommendations, a modified DR-TMP can be designed to maximize grain refinement in 7075 Al. Table VI lists parameters measured for the modified treatment, along with predicted and experimentally determined grain sizes. The modified DR-TMP addresses each recommendation described in the preceding paragraph, and provides enhanced grain refinement compared with the original 400°C/8 h overaging DR-TMP.

Although the modified DR-TMP was empirically devised before quantitative modeling results were available, it provides an example of use

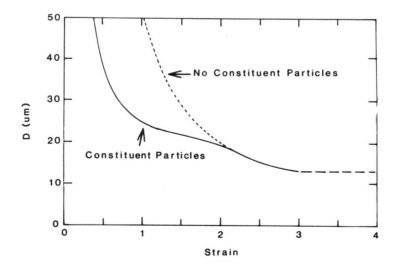

Fig. 12   Effect of constituent particles on recrystallization-limited and grain-growth-limited grain size.

TABLE VI

Modeling Parameters for Modified DR-TMP

Parameter	Original TMP	Modified TMP	Units
B	5.3×10^6	2.2×10^6	m^{-1}
C	3.8×10^{24}	1.9×10^{23}	m^{-4}
f_E	0.045	0.077	
f_D/r_D	5.0×10^4	5.0×10^4	m^{-1}
D (predicted)	15	12	μm
D(experimental)	15	11	μm

of this model to design thermomechanical processing methods for enhanced grain refinement. Extensive experimentation with modeling parameters suggests that the modified DR-TMP associated with the values shown in Table VI provides the finest grain sizes attainable in 7075 Al by DR-TMPs employing a deformation strain of 2.5 or less.

5.6   Combined Alloy/TMP Selection

Since all of the examples of TMP development described in this paper have been based on processing of 7075 Al, it is natural to ask whether other alloy/TMP combinations can promise finer stable grain sizes. It appears that two approaches are available, as described in the closing paragraphs.

5.6.1 Development of New Alloys for DR-TMPs

Estimation of dispersoid f_D/r_D values from alloy compositions shows that most commercial alloys provide f_D/r_D values between 4×10^4 and 8×10^4 m^{-1}. Modeling suggests that transition from grain-growth limited grain size to recrystallization-limited grain size occurs near $f_D/r_D = 6 \times 10^4$ m^{-1} for these alloys. Thus, these existing alloys have roughly optimum dispersoid f_D/r_D values for the volume fraction of equilibrium phase that can be precipitated.

Modeling indicates that substantially finer grain sizes cannot be achieved by DR-TMP processes, within constraints imposed by heat-treatable alloy compositions. However, this result is predicted by a model which has not been tested for values of f_D/r_D exceeding roughly 8×10^4 m^{-1}. Changes in either boundary mobility or stored deformation energy could be produced by altering alloy chemistry or dispersoid characteristics. Such changes would affect the f_D/r_D value corresponding to the transition point from grain-growth limited to recrystallization-limited grain size. Thus, experiments to examine DR-TMPs of alloys with $f_D/r_D > 10^5$ m^{-1} appear worthwhile, even though the present model suggests little further grain refinement in such alloys.

5.6.2 Development of Alloys for CR-TMP

Development of alloys for CR-TMPs appears to hold more promise for achieving stable grain sizes below 5 μm. Such alloys need to have quite high dispersoid f_D/r_D values, which are difficult to achieve using standard large-scale casting equipment. Various power-metallurgy methods appear promising for fabrication of such alloys. Unfortunately, our present understanding of continuous recrystallization is so limited that development of optimum alloy/CR-TMP combinations requires extensive experimentation. Improved understanding of continuous recrystallization, leading to models of CR-TMPs similar to the DR-TMP model outlined in the present paper, appears to be the next challenge in the quest for property control by thermomechanical processing of heat-treatable aluminum alloys.

7. Acknowledgments

The author wishes to thank Rockwell International for supporting the work reported in this paper. I express my gratitude to Prof. E. Nes for discussing many details of recrystallization modeling with me; and to C. H. Hamilton, M. W. Mahoney, A. K. Ghosh, J. C. Chesnutt, C. G. Rhodes, P. A. Wycliffe and other colleagues for critical evaluation of many aspects of this work. Invaluable experimental assistance has been provided by A. B. Ward, M. Calabrese, R. A. Spurling and P. Q. Sauers. Thanks are also extended to R. D. Doherty for helpful discussions.

8. References

1.   V.A. Philips, A.J. Swain and R. Eborall, J. Inst. of Metals, 81 (1952) 625-647.

2.   R.P. Carreker and W.R. Hibbard, Trans. AIME, 209 (1957) 1157-1163.

3.   A.W. Thompson, M.I. Baskes and W.F. Flanagan, Acta Met., 21 (1973) 1017-1028.

4.   N. Hansen, Acta Met., 25 (1977) 863-869.

5.   D.J. Lloyd, Metal Science, 14 (1980) 193-198.

6.   J.A. Wert, "The Flow Stress-Grain Size Relationship in a Precipitation Hardening Aluminum Alloy," in Sixth International Conference on the Strength of Metals and Alloys, R.C. Gifkins (ed.), Pergamon Press, Oxford, 1982, pp. 339-344.

7.   W. B. Morrison and R. L. Miller, Ultrafine-Grain Metals, J. J. Burke and V. Weiss, eds., Syracuse University Press, Syracuse, N.Y., 1970, pp. 192-211.

8.   R. J. McElroy and Z. C. Szkopiak, Inter. Metall. Rev., 17 (1972), 175-202.

9.   A. W. Thompson, Met. Trans. 8A (1977) 833-842.

10. A.K. Ghosh and C.H. Hamilton, Met. Trans., 13A (1982) 733-743.

11. R.C. Gifkins, "Mechanisms of Superplasticity," in Superplastic Forming of Structural Alloys, N.E. Paton and C.H. Hamilton (eds.), The Metallurgical Society of AIME, Warrendale, 1982, pp. 3-26.

12. D.J. Lloyd and D.M. Moore, "Aluminum Alloy Design for Superplasticity," in Superplastic Forming of Structural Alloys, N.E. Paton and C. H. Hamilton (eds.), The Metallurgical Society of AIME, Warrendale, 1982, pp. 147-172.

13. C.H. Hamilton, C.C. Bampton and N.E. Paton, "Superplasticity in High Strength Aluminum Alloys," in Superplastic Forming of Structural Alloys, N.E. Paton and C.H. Hamilton (eds.), The Metallurgical Society of AIME, Warrendale, 1982, pp. 173-189.

14. P. Lagarde and M. Biscondi, Canadian Metallurgical Quarterly, 13 (1974) 245-251.

15. M. Biscondi and C. Goux, Mem. Sci. Rev. Metall., 65 (1968) 167-179.

16. R. A. Benedek and R. D. Doherty, Scripta Met., 8 (1974) 675-678.

17. E. Hornbogen, "Design of Heterogeneous Microstructures by Recrystallization," in Fundamental Aspects of Structural Alloy Design, R.I. Jaffee and B.A. Wilcox (eds.), Plenum Press, New York, 1977, pp. 389-409.

18. T.H. Alden, "Processing and Properties of Superplastic Alloys," in Fundamental Aspects of Structural Alloy Design, R.I. Jaffee and B.A. Wilcox (eds.), Plenum Press, New York, 1977, pp. 411-430.

19. J.A. Wert, "Grain Refinement and Grain Size Control," in Superplastic Forming of Structural Alloys, N.E. Paton and C.H. Hamilton (eds.), The Metallurgical Society of AIME, Warrendale, 1982, pp. 69-83.

20. H. Alborn, E. Hornbogen and U. Koster, J. Mat. Sci., 4 (1969) 944-950.

21. J.W. Martin and R.D. Doherty, Stability of Microstructure in Metallic Systems, Cambridge University Press, Cambridge, 1976, pp. 150-153.

22. R. D. Doherty, "Nucleation," in Recrystallization of Metallic Materials, 2nd Edition, F. Haessner (ed.), Riederer-Verlag, Stuttgart, 1978, pp. 40-61.

23. E. Hornbogen and U. Koster, "Recrystallization of Two-Phase Materials," in Recrystallization of Metallic Materials, 2nd Edition, F. Haessner (ed.), Riederer-Verlag, Stuttgart, 1978, pp. 159-194.

24. F.J. Humphreys, Acta Met., 25 (1977) 1323-1344.

25. F.J. Humphreys, "Nucleation of Recrystallization in Metals and Alloys," 1st Riso Int. Symp. on Metallurgy and Materials Science, Roskilde, Denmark, 1980, pp. 35-44.

26. R. Sandstrom, "Criteria for Nucleation of Recrystallization Around Particles," 1st Riso Int. Symp. on Metallurgy and Materials Science, Roskilde, Denmark, 1980, pp. 45-49.

27. R.D. Doherty, "Nucleation of Recrystallization in Single Phase and Dispersion Hardened Polycrystalline Materials," 1st Riso Int. Symp. on Metallurgy and Materials Science, Roskilde, Denmark, 1980, pp. 57-69.

28. R.D. Doherty, Metal Science, 8 (1974) 132-142.

29. C.S. Smith, Trans AIME, 175 (1948) 15-48.

30. E. Nes, N. Ryum and O. Hunderi, Acta Met., 33 (1985) 11-22.

31. L.F. Murr, Interfacial Phenomena in Metals and Alloys, Addison-Wesley, Reading, 1975, p. 131.

32. R. Viswanathan and C.L. Bauer, Acta Met., 21 (1973) 1099-1109.

33. H. Hu, "Recovery, Recrystallization and Grain Growth," in Metallurgical Treatises, J.K. Tien and J.F. Elliot, (eds.), The Metallurgical Society of AIME, Warrendale, 1981, 385-407.

34. P. Cotterill and P.R. Mould, Recrystallization and Grain Growth in Metals, Surrey University Press, London, 1976, pp. 181-249.

35. E. Nes, "Superplastisitet og høyduktilitet i Zr-holdige aluminiumlegeringer," NTNF Report 80 01 52 -1, 1980, 26 pp.

36. R. Grimes, C. Baker, M. J. Stowell and B. M. Watts, Aluminum, 51 (1975) 720-723.

37. R. Grimes, M.J. Stowell and B.M. Watts, Metals Technology, 3 (1976) 154-160.

38. B.M. Watts, M.J. Stowell, B.L. Baikie and D.G.E. Owen, Metal Science, 10 (1976) 189-197.

39. B.M. Watts, M.J. Stowell, B.L. Baikie and D.G.E. Owen, Metal Science, 10 (1976) 198-206.

40. R. H. Bricknell and J. W. Edington, Met.Trans., 10A (1979) 1257-1263.

41. J. W. Edington, Met. Trans., 13A (1982) 703-715.

42. C.J. Tweed, B. Ralph and N. Hansen, Acta Met., 32 (1984) 1407-1414.

43. C.C. Bampton, J.A. Wert and M.W. Mahoney, Met. Trans., 13A (1982) 193-198.

44. E. Di Russo, M. Conserva, M. Buratti and F. Gatto, Mater. Sci. Eng., 14 (1974) 23-36.

45. J. Waldman, H. Sulinski and H. Markus, Met. Trans., 5 (1974) 573-584.

46. J. Waldman, H. Sulinski and H. Markus, "Thermomechanical Processing of Aluminum Alloy Ingots," in Aluminum Alloys in the Aircraft Industry, Technicopy Ltd., Glouchestershire, 1976, pp. 105-114.

47. B.K. Park and J.E. Vruggink, "Properties and Structure of Thermomechanical-Processed 7475-T6 Plate," in Thermomechanical Processing of Aluminum Alloys, T.G. Morris, ed., The Metallurgical Society of AIME, Warrendale, PA, 1979, pp. 25-49.

48. R.E. Sanders, Jr. and E.A. Starke, Jr., "The Effect of Ingot Processing on the Microstructure and Fatigue Properties of 7475-T6 Plates," in Thermomechanical Processing of Aluminum Alloys, T.G. Morris, ed., The Metallurgical Society of AIME, Warrendale, PA, 1979, pp. 50-73.

49. J.A. Wert, N.E. Paton, C.H. Hamilton and M.W. Mahoney, Met. Trans., 12A (1981) 1267-1276.

50. E. Nes and J.A. Wert, Scripta Met., 18 (1984) 1433-1438.

51. R. Sandstrom, Z. Metallk., 71 (1980) 741-751.

52. M. Avrami, J. Chem. Phys., 7 (1939) 1103-1112.

53. M. Avrami, J. Chem. Phys., 8 (1940) 212-224.

54. M. Avrami, J. Chem. Phys., 9 (1941) 177-184.

55. W.A. Johnson and R.F. Mehl, Trans. AIME, 135 (1939) 416-442.

56. E. Nes, "Grain Size and Texture Control in Commercial Aluminum Alloys," Proc. 7th Int. Light Metals Congress, Leoben/Vienna (1981) 154-155.

57. E. Nes, Metalurgia I. Odlwunictwo, 5 (1979) 209-224.

58. E. Nes, Recrystallization in Alloys with Bimodal Particle Size Distributions," 1st Riso Int. on Metallurgy and Materials Science, Roskilde, Denmark, 1980, pp. 85-95.

59. M.F. Ashby, J. Harper and J. Lewis, Trans AIME, 245 (1969) 413-420.

60. J.T. Staley, Metals Engineering Quarterly, 16, no. 2 (1976) 52-57.

61. D.S. Thompson, Met. Trans., 6A (1975) 671-683.

GRAIN SIZE AND TEXTURE CONTROL IN COMMERCIAL ALUMINIUM ALLOYS

E. Nes

The Norwegian Institute of Technology
Department of Physical Metallurgy
N-7034 TRONDHEIM-NTH, Norway

Summary

The structural parameters which control the recrystallized grain size
and the strength of the cube texture in commercial aluminium alloys have
been discussed in terms of a simple theoretical model. These parameters
are: The size distribution of the large primary particles, the stored
deformation energy, the solid solution content and the Zener drag pressure
which depends on the size and volume fraction of dispersoids. How these
parameters according to the model are playing together has been illustrated
by the construction of grain size maps for some important commercial
alloys.

1. Introduction

The following is an attempt to analyse both the grain size- and the
texture aspects of recrystallization within the framework of a single
model. Although these aspects of recrystallization are intemately connected
they will, for practical reasons, be covered in separate sections below.

2. Grain size control

The recrystallization behaviour of alloys with a bimodal particle size
distribution was modeled some years ago by the present author (1,2), with
some additional refinements given in Ref. 3. This model has resently been
further developed by Nes and Wert (4) and will in the present work be
applied in the analysis of grain structures resulting from the processing
of commercial aluminium alloys.

2.1 Model

The model rests on the assumption that nucleation of recrystallization
in an alloy with a bimodal particle size distribution is restricted to the
deformation zones surrounding large particles. To act as a potential

95

nucleation site, a particle must exceed a critical diameter given by

$$\eta_c = \frac{2\gamma}{3(P_D - P_Z)} \tag{1}$$

where γ is the grain boundary energy, P_D is the driving pressure for recrystallization due to the stored deformation energy, and P_Z is the Zener pressure due to the drag from small dispersoid particles. If P_D and P_Z can be evaluated, then the number of potential nucleation sites per unit volume, $N(\eta_c)$, can be calculated from the particle size distribution $f(\eta)$

$$N(\eta_c) = \int_{\eta_c}^{\infty} f(\eta) \, d\eta \tag{2}$$

In order to calculate the recrystallized grain size from the number of potential nucleation sites, the recrystallization kinetics needs to be established. In materials where the new grains originate at well defined sites, the Avrami theory (5) should apply, provided the nucleation sites are randomly distributed. According to Avrami's theory, there are preexisting in the matrix a limited number of potential nucleation sites ($N(\eta_c)$ in our model), each having a nucleation frequency ν. During recrystallization, the sites are used up by nucleating new grains and the number of potential sites N remaining after time t is; $N = N(\eta_c)\exp(-\nu t)$, giving the nucleation rate as $\dot{N} = N(\eta_c) \nu \exp(-\nu t)$. The recrystallization kinetics and the average recrystallized grain size can easily be evaluated in the following two extreme cases:

(i) For νt^* very large (t^* is the time for complete transformation), it follows that $\dot{N} \to 0$ for $t \ll t^*$. In this case, the recrystallization reaction is site saturated, and the average grain size is given by $D \simeq (N(\eta_c))^{-1/3}$.

(ii) For $\nu t^* \to 0$, we get the Johnson-Mehl case (6) with $\dot{N} = \nu N(\eta_c)$. In this case, the average recrystallized grain size will be

$$D = \left(\frac{G}{\nu N(\eta_c)} \right)^{1/4} \tag{3}$$

where G is the growth rate of a grain growing into the untransformed matrix (assumed to be constant).

Observations in Al-Mn alloys (3) indicate that the Johnson-Mehl case is the most realistic of the two. The growth rate G can be written $G = M(P_D - P_Z)$, where M is the grain boundary mobility. Combining this expression for the growth rate with Eq. (3) gives

$$D = K \left[\frac{P_D - P_Z}{N(\eta_c)} \right]^{1/4} \tag{4}$$

where $K = (M/\nu)^{1/4}$ is assumed to be a constant for a constant recrystallization temperature. As a first order approximation, we do not expect the mobility M to vary with P_D or P_Z. The nucleation frequency, ν, however, may vary with the amount of cold reduction (i.e., with P_D). However, for strains larger than 1, the structure within the deformation zones approaches saturation, so a constant nucleation frequency may also be a reasonable first order approximation, at least for large strains.

2.2 Application

This model will now be applied in the analysis of the recrystallization behaviour of some commercial non-heat-treatable wrought aluminium
alloys studied by Andersson (7) and the present author (2). This alloy
selection includes three DC-cast 1100-series alloys (Al0.6Fe0.5Si,
Al0.2Fe0.2Si and Al0.5Fe0.14Si), a strip cast Al0.5Fe0.2Si alloy, a DC-cast
Al1.2Mn alloy and a strip cast and homogenized Al0.8Mn alloy.

It follows from Eq. (4) that in order to predict the recrystallized
grain size, the following three parameters need to be established: 1) the
size distribution of the large particles; 2) the stored deformation energy;
and 3) the Zener drag pressure which depends on the size and volume
fraction of dispersoids.

Particle Size Distributions. The particle size distribution for the
three DC-cast 1100-series alloys are given in Fig. 1. The total number of
particles per unit volume $N(\eta)$ with diameter larger than η has been
derived from the two-dimensional experimental distributions, i.e. the
number of particles $N(d)$ per unit area in a polished surface which are
larger than d. The 2D $\rightarrow$ 3D transformation is based on the Sandström
analysis (9). Corresponding size distributions for the AlMn alloys can be
found in Ref. 3.

The Driving Pressure. As a first order estimate, the driving pressure
for the recrystallization front is assumed to be

$$P_D = \rho \tau \qquad (5)$$

where ρ is the dislocation density and τ the dislocation line tension
($\tau = \frac{1}{2}\mu b^2$ where μ is the shear modulus and b the Burger's vector). The
dislocation density can be estimated from the mean subgrain size, d, and
the mean subgrain misorientation, θ .

$$\rho = \kappa \frac{\theta}{b\,d} \qquad (6)$$

where κ is a constant of order unity.

The variation in subgrain size with rolling strain in a 1100-series
alloy has been studied in a previous investigation (10) with the results
given in Fig. 2.

The mean subgrain size can be estimated as d = 1/3 (2 d_L + d_S), where
L denotes longitudinal and S denotes short transverse. It has been found
that the misorientation, θ , is rather insensitive to strain for ε > 0.5.
During rolling this misorientation rapidly saturates at about 2° (10).
Combining Eqs. (5) and (6) gives

$$P_D = 0.13 \frac{\kappa}{d} (N\ m^{-2}) \qquad (7)$$

(In deriving this, μ = 2.7 10^{10}N/m^2 and b = 2.86 10^{-10}m have been used).

The Zener Drag. The Zener drag pressure can be written in the form

$$P_Z = \alpha \gamma \frac{f}{r} \qquad (8)$$

where γ is the grain boundary energy (γ - 0.3 N/m in Al, (11)), f is the
volume fraction of dispersoids, r is the mean dispersoid radius and α is a
constant of order unity. We will use α = 3/4 which is the value derived by

97

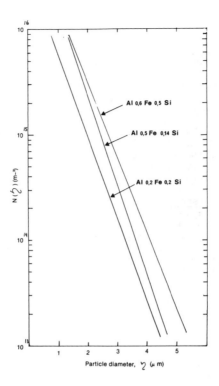

Figure 1 - The number of particles per unit volume $N(\eta)$ of size larger than η for the three alloys identified in the diagram. The curves are derived from observations given in Ref. 8.

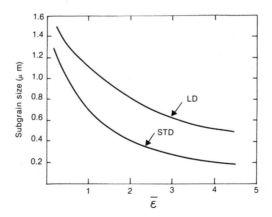

Figure 2 - Subgrain size as a function of rolling reduction in a commercial 1100 aluminium alloy. STD and LD refer to short transverse and the rolling directions respectively. Data taken from Ref. 10.

Zener (12). This value of α is as reasonable as any other α-values derived later (13).

Evaluation of Constants. The constants K and κ in the Eqs. 4 and 7 which give the best fit to the experimental results were $K = 6.5 \; 10^{-3}$ $(m/N)^{1/4}$ and $\kappa = 0.39N/m$. These values correspond to a critical particle size for nucleation of recrystallization of $\eta_c = 2\mu m$ at a rolling reduction of about $\varepsilon = 3$.

The model predictions and the experimental results are given in Figs. 3 and 4. Figure 3a and b refer to DC- cast aluminium manganese. The model and experiments are in reasonably good agreement, although in the case of the strip cast alloy the model underestimates the recrystallized grain sizes for lower rolling reductions. Figure 4 shows the results for the DC-cast 1100 series alloys. The theoretical curves are in good agreement with the observations.

In analysing the results given in Figs. 3 and 4 the Zener drag due to the dispersoids have been neglected, i.e. $P_z = 0$ in Eq. 4. The results given in Fig. 5, however, showing the grain size variation with strain in the surface- and in the central regions of the strip cast and homogenized Al0.5Fe0.2Si alloy are interesting in terms of a P_z effect. Note that the

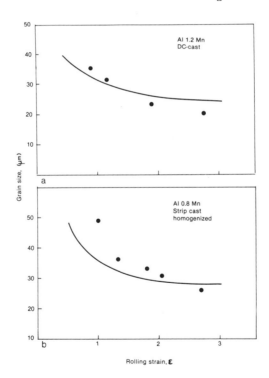

Figure 3 – Recrystallized grain size as a function of rolling reduction for
a) DC-cast AlMn and b) strip cast and homogenized AlMn alloy
(Ref. 3). The fully drawn curves are theoretical predictions.

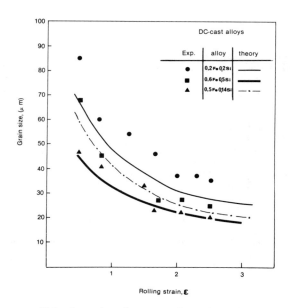

Figure 4 – Recrystallized grain size as a function of rolling strain. The experimental data and theoretical curves are identified in the diagram. Experimental data taken from Ref. 7.

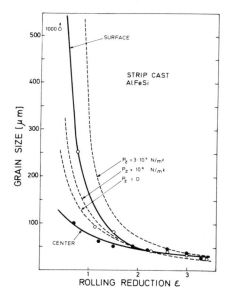

Figure 5 – Recrystallized grain size in the surface region and in the center as a function of rolling reduction for a strip cast and homogenized 1100 alloy. The theoretical curves (broken) correspond to the Zener pressure values given. Experimental data taken from Ref. 7.

recrystallized grain size in the surface at low strains is about an order of magnitude larger than in the central region (1000µm vs 100µm), while at large strains the grain size is about equal in the two regions. As shown by the theoretical curves in Fig. 5 this difference can be explained as due to the higher density of dispersoids in the surface region. The dispersoid density in the center amounts to a volume fraction of about 0.03 of particles with an average radius of about 0.5µm, which according to Eq. 8 corresponds to a Zener drag $P_Z = 10^4$ N/m^2. As shown in Fig. 5 such a Zener pressure has only a small effect on the recrystallized grain size as compared to the $P_Z = 0$ case. However, increasing P_Z to $3 \cdot 10^4$ N/m^2 will have a significant impact on the recrystallization behaviour. As the density of dispersoids are significantly higher in the surface region this may explain the exceedingly large surface grain size at low strain in this case.

2.3 Effects of Elements in Solid Solution

It has been suggested by the present author (1) that a possible effect of elements in solid solution on the recrystallized grain size is a grain refinement due to the larger amount of stored dislocations. The stored deformation energy P_D at a given ε will increase with increasing amount of alloying elements in solution and accordingly the grain size will decrease as predicted by Eq. 4. However, this is only so as long as the amount in solid solution in the cold rolled state is still in solid solution at the onset of recrystallization. If precipitation occurs , the benefical effect on P_D may rapidly be counter-balanced by an increase in P_Z. Assuming that the elements in solution remain so during the recrystallization treatment, there is an added effect on the driving pressure P_D also due to less amount of recovery occuring before the nucleation of new grains. A nice example to illustrate this solid solution effect on the recrystallized grain size is the effect of small Mg-additions on the recrystallized structure in 3000-series alloys (14).

2.4 Grain Size Maps

The simple theory above provides a good tool for a quantitative prediction of the grain size in commercial aluminium sheet alloys after normal production thermomechanical treatments. As stated above, the salient parameters controlling the grain size are: (i) The size distribution of large particles; (ii) the stored deformation energy; and (iii) the Zener drag pressure. From a sheet metal producer's point of view, the concern is how to obtain a grain size which is better than some critical value, or a grain size which does not exceed a critical value, say in the range 50-70µm. How to meet such a demand in a certain alloy system can be conveniently illustrated by the construction of grain size maps as shown in Fig. 6. These maps are easily generated from the theory. The maps divide the Zener drag rolling strain (P_Z/ε)-space into a coarse-grained region (grain size larger than 60µm) and a fine-grained region (grain size smaller than 30µm). The position of the dividing boundary is determined by the large particle size distribution. The maps in Fig. 6 span the whole regime of commercial aluminium ingot alloys, from strip cast 1100 to DC-cast 7075. The map in Fig. 6a, shows that strip cast AlMn cannot be produced with a grain size smaller than about 100µm. On the other side, the highly alloyed 7075 can be produced with a grain size less than 10µm (Fig. 6d), i.e. a material with superplastic properties. This map (Fig. 6d) has been constructed based on the observations and analysis by Wert et al (15) and Nes and Wert (4). DC cast AlMn alloys (Fig. 6c) can be produced without any grain size problems in conventional sheet. However, can-stock production may be associated with grain size problems. A high dispersoid density will have a beneficial effect on texture, but increasing P_Z will have a negative effect on grain size.

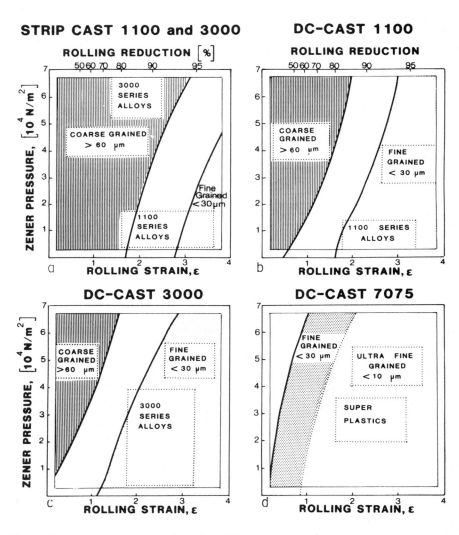

Figure 6 – Grain size maps for the alloys given. Combinations of Zener pressure values and rolling reductions which are typical of commercial sheet production are indicated in the diagrams.

3. Texture Control

Before discussing the basic mechanisms controlling the recrystallization texture development in commercial aluminium alloys, it may be useful to look at some general observations regarding how the strength of the cube texture component is influenced by variations in: the particle content; elements in solid solution; rolling temperatures; etc. Regarding the general effect of alloy composition on the cube strength, experience shows a relationship as outlined in Fig. 7. In super purity material recrystallization gives nearly 100% cube texture, while in commercial alloys the cube strength varies in the range from a few % up to 20-30%. However, it is possible to subject alloys of commercial composition to special thermomechanical treatments which may give much higher volume fraction of cube texture after recrystallization. Therefore the variation in cube strength* with composition is given as a rather broad scatterband in Fig. 7. Materials containing an alloy addition of around 1wt%, i.e. the 1100- and 3000-series alloys can easily be processed to give cube volume fractions in the range from about 0% to 50%. The most effective parameters to play with in this regard are:

(i)   The temperature of rolling. As shown in Fig. 8 the cube/R ratio may vary from 0.2 to 1 by increasing the rolling temperature from liquid nitrogen temperature to 100°C (16). Further, it is well established that hot rolling may result in cube volume fractions of 50% or more.

(ii)   The dispersoid content. The strength of the cube texture component in commercial alloys will increase with increasing dispersoid density as has been demonstrated in the case of AlMn alloys (2).

(iii) The amount of Fe-Si in solid solution in the cold rolled state may have a profound effect on the volume fraction of the cube or the cube/R ratio after recrystallization.

3.1 Mechanisms

The following is an attempt both to interpret the general form of the curve in Fig. 7 and to establish the mechanisms responsible for the wide scatterband, i.e. the mechanisms behind the temperature-, dispersoid-, and solid solution effects. The analysis rests on the following observations/assumptions regarding the recrystallization behaviour of commercial aluminium alloys:

(i)   The major volume fraction ($\geq$70%) of the fully recrystallized material is due to particle stimulated nucleation of recrystallization. The analysis in the first part of this paper supports this view, and direct evidence for the importance of particles in the nucleation of recrystallization in 1100-series alloys has been given by Bay and Hansen (17).

(ii) Grains growing out from particle sites contribute to a R-, random texture. This is documented in the case of 1100-series alloys by the investigations by Juul Jensen et al (18) and Nes and Solberg (19) and for the 3000-series alloys by Anderson and Nes (20).

---

*The curve in Fig. 7 is drawn to illustrate a general trend. Large deviations from this may occur. For instance, if specially textured material or single crystals are used in the rolling and annealing experiment a zero volume fraction of cube texture will be obtained even in superpurity metals.

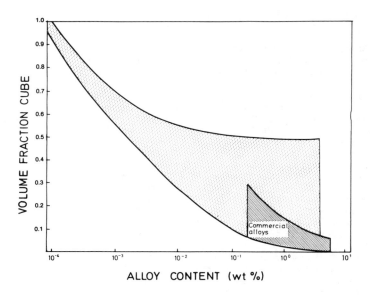

Figure 7 – The diagram illustrates the general trend in the variation of
the strength of the cube texture component with alloy content.
The dotted area reflects scatter in the results.

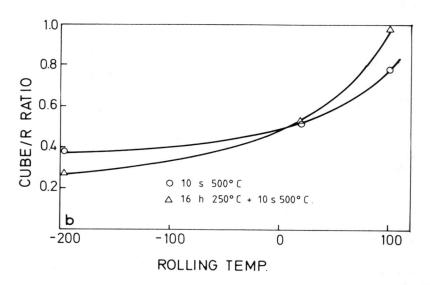

Figure 8 – The variation in the cube/R-ratio with rolling temperature (16).
The two curves refer to the different recrystallization heat
treatments given in the figure.

104

(iii) The cube grains are due to special sites (no particles involved) found only in material subjected to plane strain deformation.

(iv) The nucleation rate of cube oriented grains is less influenced by the presence of dispersoids, i.e. the Zener pressure, as compared to particle deformation zone nuclei.

(v) The volume fraction of cube oriented material increases relative to the other texture components during the recrystallization transformation. Direct observation of this effect in 1100-series alloys is given in the recent work by Juul Jensen et al (18) and Nes and Solberg (19).

The points (iii) and (iv) require some additional comments. Point (iii) is based on the assumption that the cube nucleation sites in commercial aluminium alloys are similar to those identified in cold rolled copper by Ridha and Hutchinson (21), i.e. transition band heterogeneities of the type described by Dillamore and Katoh (22). In support of this view are the recent observations of such heterogeneities in cold rolled superpurity aluminium (23) and 1100-series alloys (16). Regarding point (iv) above, it follows from the special morphology of the cube sites, being bands of 0.5-1μm in thickness and 10-20μm long (and wide) (21,23) and separated from the surrounding matrix by high angle boundaries that the energy barrier for growth of recrystallization out of such a cube site should be very small.

It follows from these assumptions that the final recrystallization texture is a compromise which reflects: Firstly, differences in the number of nucleation events which belong to the different texture components; and secondly, differences in the transformation behavior which may be due either to the spatial distribution of the different type sites, or to differences in the growth rates of grains of the different texture components. On the general form of the curve in Fig. 7, this is most likely explained in terms of the nucleation rate effect. By increasing the alloy content this will cause an increase in the number of particle sites, $N(\eta_c)$. If $N(\eta_c)$ increases, then the fraction of cube sites, $N(cube)/N(\eta_c)$ will decrease which again will result in a reduced volume fraction of cube oriented material. But as stated in point (v) above the volume fraction of cube oriented material increases relative to the other components during the transformation, $f(cube)/f(other) > N(cube)/N(\eta_c)$. As discussed in detail in Ref. 19 this is attributed to the spatial distribution of the cube sites which gives the growing cube grains an advantage in the sense of being more free to grow without impinging on neighbour grains. No evidence of differences in growth rates was reported in Ref. 19. Accordingly, the increase in the volume fraction of cube during the transformation period will be proportional to the ratio L_c/L_p where L_c and L_p refer to the average next neighbour distance measured from a cube site and a particle site, respectively. In conclusion, the volume fraction cube oriented material in the fully recrystallized alloy should be a function of two ratio-parameters, i.e.

$$f_v^{cube} = f(\frac{N(cube)}{N(\eta_c)} , \frac{L_c}{L_p}) \qquad (9)$$

It follows from this that the final texture reflects both the nucleation and growth aspects of the transformation. How these two parameters interact is difficult to answer. However, as a first order estimate the strength of the cube-component is likely to be a function of the number of particle sites as shown schematically in Fig. 9. Decreasing

105

the number of particle sites will favour the cube-sites and it follows that the strength of the cube will follow the grain size. Therefore the grain size model above can be applied as a tool to give qualitative predictions on the strength of the cube texture. Some important consequences are given below:

On the effect of rolling temperature: By increasing the rolling temperature the stored deformation energy, P_D will decrease due to dynamic recovery. A decrease in P_D will according to Eq. 2 result in an decrease in $N(\eta_c)$ which according to Eqs. 4 and 9 will result in an increase in grain size and in the volume fraction of cube oriented material respectively. This effect is clearly illustrated in Table 1, which compares the grain size and cube strength of some 1100-series alloys after cold rolling and hot rolling. The results in Table 1 are taken from Ref. 16.

Table 1. Grain size and cube strength in three DC-cast 1100 series alloys after hot rolling and after 95% cold rolling and annealing for 10s at 510°C. (Ref. 15).

| Condition | Al0.6Fe0.5Si | | Al0.5Fe0.14Si | | Al0.3Fe0.05Si | |
	Grain size µm	Cube strength X Random	Grain size µm	Cube strength X Random	Grain size µm	Cube strength X Random
Cold rolled	34	3	27	5	28	10
Hot rolled	140	9	170	10	115	18

On the effect of dispersoids : By increasing the Zener drag, P_z, this will according to the model result in increasing grain size and this grain size effect is paralleled by an increase in the strength of the cube component in DC-cast AlMn alloys (2, 18).

On the effect of elements in solid solution: There are few fields in physical metallurgy where there are so much conflicting interpretations and experimental observations as on the effect of Fe and Si on recrystallization texture in 1100-series aluminium. One reason for this is that the Fe/Si-content will influence the texture development in rather different directions depending on whether these elements are present as a solid solution or precipitated as finely dispersed particles. Even if Fe and Si prior to the recrystallization annealing treatment is in solid solution, the chanses are that these elements may precipitate prior to the nucleation of recrystallization, this is likely to occur especially during batch annealing treatments. As mentioned above, finely dispersed particles promote the cube texture. If care is taken to keep Fe-Si in solid solution during the transformation the primary effect of these elements is to increase the amount of stored deformation energy during rolling, i.e. to increase the driving pressure P_D for recrystallization which in turn stimulates particle nucleation ($N(\eta_c)$ increases) and the grain size and cube strength decreases.

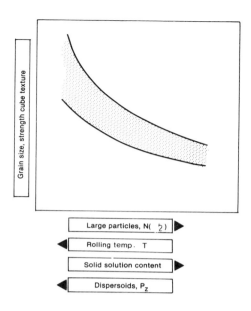

Figure 9 – The recrystallized grain size and the strength of the cube texture component as a function of the parameters given.

4. Acknowledgments

The author wants to thank BFMT (Germany), NTNF (Norway), Vereignigte Aluminium-Werke AG and Årdal og Sunndal Verk A/S for financial support.

5. References

1. E. Nes, Metalurgia I. Odlwunictwo, 5 (1979) 209-224.

2. E. Nes, "Recrystallization in Alloys with Bimodal Particle Size Distributions", 1st. Risø Int. Symposium on Metallurgy and Materials Science, Roskilde, Denmark 1980, pp. 85-95.

3. E. Nes, Proc. of 7th Int. Light Metal Congress, Leoben/Vienna (1981) pp. 154-155.

4. E. Nes and J.A. Wert, Scripta Met., 18 (1984) (1433-1438).

5. M. Avrami, J. Chem. Phys., 7 (1939) 1103-1112; 8 (1940) 212-224; 9 (1941) 177-184.

6. W.A. Johnson and R.R. Mehl, Trans. AIME, 135. (1939) 416-442.

7. B. Anderson, "Grain size variation in AlFeSi alloys". Si-Report No. 82 0160-2 (1982) pp. 18.

8. Y. Langsrud, "Size Distribution of Particles in DC-cast AlFeSi Alloys", SI-Report No 810142-1 (1983) pp. 17.

9. R. Sandstrøm, Z. Metallk. 71, 1980 741-751.

10. E. Nes, A.L. Dons and N. Ryum, "Substructure Strengthening of Cold Rolled Aluminium Alloys", Proc. 6th Int. Conference on Strength of Metals and Alloys. (ICSMA6), Pergamon Press, New York, 1982 pp. 425-430.

11. L.F. Murr, Interfacial Phenomena in Metals and Alloys, Addison-Wesley, Reading, 1975, p. 131.

12. C.S. Smith, Trans. AIME, 175 (1948) 15-48.

13. E. Nes, N. Ryum and O. Hunderi, Acta Met., 33 (1985) 11-22.

14. D. Altenpohl, Aluminium und Aluminium-legierungen, Springer-Verlag, Berlin, 1965 p. 699.

15. J.A. Wert, N.E. Paton, C.H. Hamilton and M.W. Mahoney, Met. Trans., 12A (1981) 1267-1276.

16. A.L. Dons and E. Nes, Materials Sci. and Tech., in press.

17. B. Bay and N. Hansen, Met. Trans. 10A, (1979) 279.

18. D. Juul Jensen, N. Hansen and F.J. Humphreys, Proc. of 7th Int. Conf. on Textures of Materials, Netherland Society for Materials, Holland, 1984 pp. 251-256.

19. E. Nes and J.K. Solberg, "On the Growth of Cube Grains During Recrystallization of Aluminium". SINTEF report STF34 F83111, 1984 pp 18

20. B. Andersson and E. Nes, "Nucleation of Recrystallization in a Commercial AlMn Alloy". 1st Ris Int. Symp. on Metallurgy and Materials Science, Roskilde, Denmark, 1980 pp. 115-119.

21. A.A. Ridha and W.B. Hutchinson, Acta Met. 30, (1982) 1929.

22. I.L. Dillamore and H. Katoh, Met. Sci., 8, (1974) 73-83.

23. E. Nes, J. Hirsch and K. Lücke, Proc. 7th Int. Conf. on Textures in Materials, Netherland Society for Materials, Holland, 1984 pp 663-668.

PLASTIC INSTABILITY DURING TENSILE DEFORMATION

OF A FINE GRAINED Al - 2wt%Fe - 0.8wt%Mn ALLOY

Håkon Westengen

Årdal og Sunndal Verk a.s., N-6600 Sunndalsøra, Norway

Abstract

Ultrafine grained sheet materials are processed from a continuously strip cast Al-2wt%Fe-0.8wt%Mn alloy. By varying the process parameters, particle stabilized grain structures with mean grain diameters in the range 0.8 - 5μm are obtained. Refinement of the grain structure provides an efficient hardening mechanism. The practical application of this hardening mechanism is, however, limited by a tendency for inhomogeneous deformation. During tensile testing, yield point effects, Lüdering and strain localization are observed. The influence of grain size, test temperature and strain rate on the tensile properties is described. It is shown that dynamic recovery effects play an important role during deformation of these materials.

Introduction

A strong plastic instability is reported to limit the tensile ductility of various fine grained ferritic steels (1-4). The phenomenon is characterized by a nominal stress vs. elongation curve showing a continuous decrease in stress after an initial yield drop. Formation of a Lüders band at the yield point is followed by further instability in the yielded zone, causing failure at low total elongations. It is generally agreed that the instability in the fine-grained steels is favoured by a decreasing grain size, decreasing testing temperature and increasing strain rate. The onset of this instability at small grain sizes imposes a limit to the practical application of grain boundary hardening in ferritic steels (1).

During the development of particle stabilized, fine-grained aluminium sheet materials based on relatively rapidly solidified low eutectic alloys, similar instability effects were observed for grain sizes below approximately 1 μm (5-6). It has further been shown that the tensile ductility of temper annealed commercial purity aluminium, AA1100, with a well developed subgrain structure is limited by strain localization following a yield drop (7). The phenomenon is also believed to be responsible for the pronounced minimum in tensile ductility as observed for the "microalloy" Al-0.5wt%Co-0.5wt%Fe for electrical conductor applications after annealing the wire at approximately 200°C (8).

In a recent study of the tensile properties of a fine-grained Al-0.8wt% Mn-2wt%Fe-alloy (9), it was noted that the instability is enhanced by a decreasing grain size. However, in contrast to the behaviour of steels, a decrease in testing temperature or an increase in strain rate tend to stabilize the deformation. These aspects are further discussed in the present work.

Experimental

The fine-grained material was processed from a continuously strip cast Al-2wt%Fe-0.8wt%Mn-alloy, following the procedure developed by Morris (6). Due to the high solidification rates ($\sim 10^3$ K/s) the eutectic reaction is partly suppressed, leaving an as cast structure consisting of primary aluminium dendrites with spacing 4-6μm, surrounded by a eutectic network. The intermetallic constituents were identified by electron diffraction (JEOL 100C) as a mixture of the cubic α-Al(Mn,Fe)Si phase and the metastable tetragonal Al_mFe. Electrical resistivity measurements (Sigmascope) indicate that approximately 0.8wt%(Mn+Fe), mainly Mn, is present in solid solution after casting. The as cast strip was annealed for 3 h at 500°C to transform and spheroidize the eutectic constituents, as well as precipitate the supersaturation of Mn and Fe. A mixture of the tetragonal Al_6(Mn,Fe) and α-Al(Mn, Fe)Si-particles with typical size 0.2-1μm dia. is formed from the eutectic constituents. The supersaturation of Mn and Fe is partly precipitated on existing particles, and partly as a new set of α-Al(Mn,Fe)Si-particles with size 0.05-0.2μm dia. About 0.15wt%(Mn+Fe) is left in solid solution after the anneal. Thus the material contains a bimodal distribution of particles with a total volume fraction of particles around 9%. The strip was then cold rolled to a final thickness 1mm corresponding to approximately 85% reduction in thickness. By selecting the annealing time and temperature during a final anneal, materials with grain sizes down to less than 1μm could be obtained.

Tensile specimens with a width 10mm and gauge length 25mm were machined prior to the final anneal and tested in a screwdriven Instron 100 kN machine, at various crosshead speeds and testing temperatures. Values of the work hardening exponent, n, were estimated by fitting the data to an equation of the form

$$\sigma = k \; \varepsilon^n$$ eq. 1.

where σ is the true flow stress, k a constant and ε the true total strain.

The microstructures were observed by transmission electron microscopy (JEOL 100C) and grain sizes were measured from scanning transmission electron micrographs. The grain size is denoted d where d equals the square root of the mean grain area. A minimum of 100 grains were measured for each sample.

Results

Effect of grain size. Nominal stress vs. elongation curves at constant testing temperature, 298K, and crosshead speed 1mm/min corresponding to an initial strain rate $\dot{\varepsilon} = 6.7 \times 10^{-4} s^{-1}$, are shown in fig. 1.

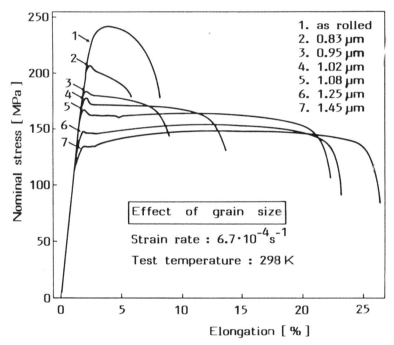

Figure 1. Load-elongation curves at various grain sizes, constant initial strain rate and test temperature.

The various grain sizes were obtained by annealing for different times at 350°C. There is obviously a critical grain size around 1.05μm below which the nominal stress decreases continuously after the initial yield drop. It is also remarkable that there is a decrease in tensile ductility during the initial period of annealing compared with the as rolled material. This occurs despite the fact that the strength is lowered significantly. As can be seen from the curves, there is a significant plastic deformation prior to the yield drop, typically around 0.5%. True uniform strain, n-values, Lüders elongation and total elongation to fracture are shown in fig. 2.

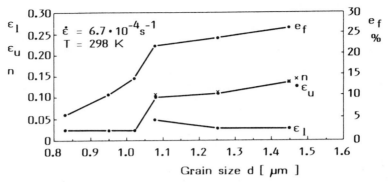

Figure 2. The influence of grain size on uniform strain ε_u, total strain to fracture e_f, lüders strain ε_L and work hardening coefficient n.

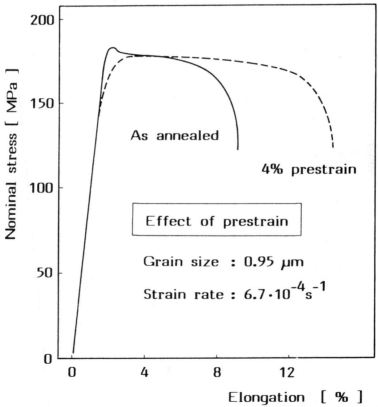

Figure 3. Load-elongation curves for as annealed material (full line) and after 4% prestrain by rolling (dotted line).

The uniform strain is taken as the strain prior to the yield drop when the nominal stress is continuously decreasing after the yield drop, and as the strain to the maximum load of the work hardening part of the curve when deformation is stable after Lüders deformation. The gradual increase of the total elongation as the critical grain size is approached from below is mainly due to the fact that an increasing part of the gauge length is traversed by the propagating Lüders bands (7). There is a remarkably good correlation between the uniform strain and the strain hardening exponent n.

If the samples with subcritical grain sizes are given a slight prestrain by rolling, the initial flow stress is lowered, the yield drop disappears, and the instability is removed, as shown in fig. 3.

Effect of testing temperature. The nominal stress vs. elongation curves for various testing temperatures at constant grain size ($\sim$1.1μm) and crosshead speed 1mm/min ($\dot{\varepsilon} = 6.7 \times 10^{-4} s^{-1}$) are shown in fig. 4.

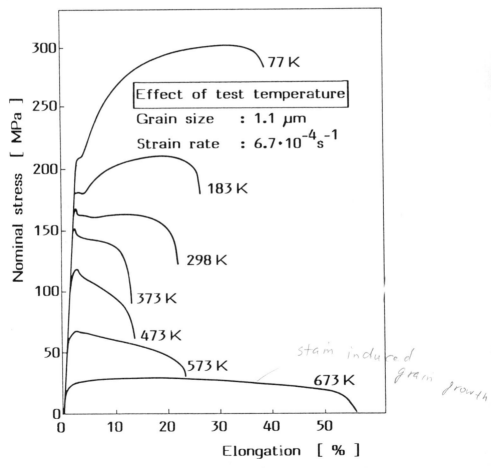

Figure 4. Load-elongation curves at various test temperatures, constant grain size and initial strain rate.

It can be seen that instability occurs at test temperatures just above room temperature. Fig. 5 shows the temperature dependence of uniform strain, total elongation to fracture and n-values. As shown elsewhere (9), the ductility increase at temperatures exceeding 600K is due to an increasing strain rate sensitivity of the flow stress. The apparent work hardening at testing temperature 673K, fig. 4, is due to strain induced grain growth (10).

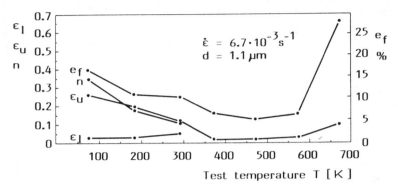

Figure 5. The influence of test temperature on uniform strain, total strain to fracture, lüders strain and work hardening coefficient.

An interesting manifestation of the influence of test temperature is shown in fig. 6.

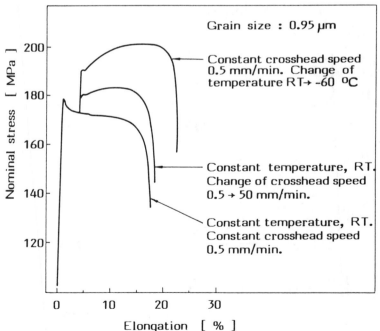

Figure 6. The effect of a change of test temperature and cross-head speed at 5% elongation, on the load-elongation curves. Note the change both in flow stress and work hardening rate.

If a sample with grain size .95μm is tested at room temperature at a strain rate 6.7 x 10^{-4} s^{-1} it is plastically unstable after the yield drop. However, if the test temperature is lowered after 5% elongation, the deformation is stabilized despite the onset of necking during the initial room temperature deformation.

Effect of strain rate. In general, an increase in the strain rate is expected to have similar effects as a decrease in testing temperature. This is indeed observed, as shown by comparing figs. 4 and 7. The nominal stress vs. elongation curves in fig. 7 were recorded at room temperature and the grain size was kept constant at approximately 1.1μm. Uniform strain, total elongation and n-values are plotted as functions of strain rate in fig. 8.

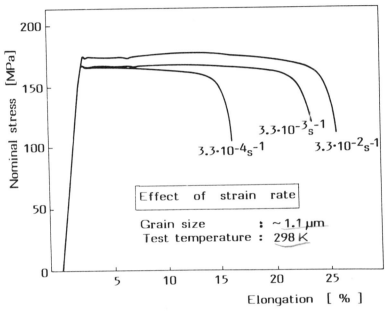

Figure 7. Load-elongation curves at various strain rates, constant grain size and test temperature.

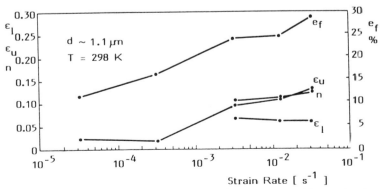

Figure 8. The influence of initial strain rate on uniform strain, total strain to fracture, lüders strain and work hardening coefficient.

The similar effects of an increase in strain rate and a decrease in test temperature is further illustrated in fig. 6. Again a sample with grain size 0.95µm is strained 5% at an initial strain rate $6.7 \times 10^{-4} \text{ s}^{-1}$, being unstable. A sudden increase in the strain rate by a factor of 100 stabilizes the deformation in the same way as a drop in the test temperature.

Strain rate change tests. The temperature and strain rate dependence of the instability indicate that dynamic recovery effects play a significant role (11). To get more detailed information on this aspect, an analysis of the strain rate dependence of the flow stress as measured by continuous tests and strain rate jump tests were performed. The material chosen for these tests was annealed for 3 hrs at 550°C followed by 3 hrs at 350°C to a grain size around 2.5µm and approximately 0.12wt%Mn in solid solution. At this relatively large grain size, the inhomogeneous yield is nearly eliminated and work hardening is appreciable, allowing the strain rate changes to be made at a range of flow stresses. To increase this range beyond the flow stresses obtained during the tensile test, a series of samples were rolled to various degrees of cold deformation. Strain rate change tests were then performed after approximately 2% tensile deformation of the predeformed specimens. Following the Cottrell-Stokes analysis (12) the parameters of interest are indicated in fig. 9.

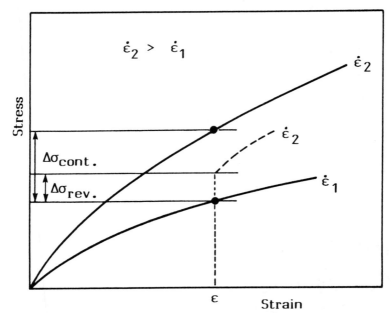

Figure 9. Schematical diagram defining the difference in flow stress during continuous tests at different strain rates $\Delta\sigma_{\text{cont.}}$, and the reversible change of flow stress during a strain rate change test $\Delta\sigma_{\text{rev.}}$

The difference in flow stress $\Delta\sigma_{\text{cont.}}$ as measured from continuous tests on different samples at different strain rates include both the contribution from the differences in structure, $\Delta\sigma_{\text{str.}}$, and the reversible change of flow stress at constant structure, $\Delta\sigma_{\text{rev.}}$, as measured by the strain rate jump test. Fig. 10. shows the variation of $(\Delta\sigma/\Delta\ln\dot{\varepsilon})_{\text{cont.}}$ and $(\Delta\sigma/\Delta\ln\dot{\varepsilon})_{\text{rev.}}$ with flow stress.

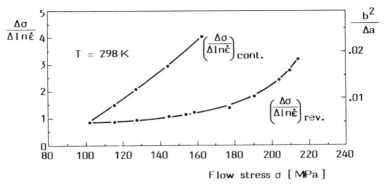

Figure 10. Haasen-plot showing the dependence of $(\Delta\sigma/\Delta\ln\dot{\varepsilon})_{rev.}$ as a function of flow stress at room temperature. Also included is the variation of $(\Delta\sigma/\Delta\ln\dot{\varepsilon})_{cont.}$

The tests were performed at room temperature and the strain rate was increased by a factor of 10 in the jump tests, base strain rate 6.7 x $10^{-4}s^{-1}$. The flow stress after the jump was determined by extrapolating the stress strain curve through the yield point back to the elastic loading line. Similar results were also obtained at 77K, fig. 11.

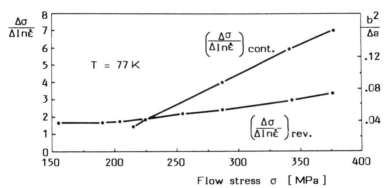

Figure 11. As fig. 10 for tests at temperature 77K.

Discussion

Onset of the instability. Following Backofen (13), the net rate of material-based strain-induced hardening $\frac{d\sigma}{d\varepsilon}$ can be written as

$$\frac{d\sigma}{d\varepsilon} = \frac{\partial\sigma}{\partial\varepsilon} + \left(\frac{\partial\sigma}{\partial\dot{\varepsilon}}\frac{\partial\dot{\varepsilon}}{\partial\varepsilon}\right) + \left(\frac{\partial\sigma}{\partial N}\frac{\partial N}{\partial\varepsilon}\right) + \ldots \qquad \text{eq. 2}$$

where N is the density of mobile dislocations. Eq. 2 indicates that there might be other factors contributing to the hardening rate.

According to Dillamore et.al. (14) the condition for shear instability is

$$\frac{d\sigma}{d\varepsilon} < 0 \qquad \text{eq. 3}$$

In the present material, the first two terms on the right side of eq. 2 are positive. Disregarding other possibilities, formation of the instabilities must therefore be due to the term involving the structural parameter N.

Annealed ultrafine grained Al-alloys contain few mobile dislocations (15). The significant yield drop which is observed after some plastic deformation indicates that a sudden increase in the value of N occurs. As discussed by Lloyd (15) this could be due to the activation of grain boundary dislocation sources, creating avalanches of dislocations. The stress concentrations associated with these avalanches activate sources in adjacent grain boundaries, causing the Lüders front to propagate. This model can easily explain the observations in fig. 3. A slight predeformation by rolling is seen to lower the initial flow stress as well as causing the yield drop to disappear. The effect of prestraining is to activate a large number of dislocation sources, which can be operated at a lower stress during the following deformation.

Development of the instability. After the Lüders band is formed at the yield drop, the further development of the instability depends upon the relative magnitude of the Lüders strain ε_L and the work hardening exponent n. The stress vs. strain curves are well described by eq. 1 for strains exceeding the Lüders strain. If we assume the same relationship to be valid for straining during lüdering, a condition for stabilization of the deformation is that (1):

$$\varepsilon_L < n \qquad\qquad\qquad \text{eq. 4}$$

The Lüders fronts will then propagate to the ends of the gauge length, followed by continued work hardening. If $\varepsilon_L > n$, necking follows initiation of the Lüders deformation, and no propagation of the Lüders band will occur. In the critical case $\varepsilon_L = n$, necking competes with the band propagation.

A relation between the parameters n, k, ε_L and the nominal lower yield stress S_{Lyp} follows by equating the Lüders stress and the stress developed by work hardening through the Lüders strain:

$$\sigma_{Lyp} = k\, \varepsilon_L^{\,n} = S_{Lyp}\,(1+e_L) \qquad\qquad \text{eq. 5}$$

where e_L is the nominal Lüders elongation. The relation can be written (1):

$$\ln \frac{k}{S_{Lyp}} = \varepsilon_L - n \ln \varepsilon_L \qquad\qquad \text{eq. 6}$$

Since the deformation is unstable if $\varepsilon_L > n$, the following stability criterion results:

$$\ln \frac{k}{S_{Lyp}} > n\,(1-\ln n) \qquad\qquad \text{eq. 7}$$

This relation is plotted as a solid curve in fig. 12.

For combinations of k/S_{Lyp} and n above the solid curve, deformation is stable, while it is unstable for values below the curve. Although the relation seems to indicate that instability is favoured at high values of n, this is not the case, as an increase in the value of n is more than compensated for by an accompanying increase in the value of k/S_{Lyp}.

The parameters n and k can only be evaluated when the deformation is stable, since they result from fitting the work hardening part of the stress vs. strain relation to eq. 1. When k/S_{Lyp} are plotted vs. n in fig. 12 it follows by extrapolation that the observed instabilities are predicted. The

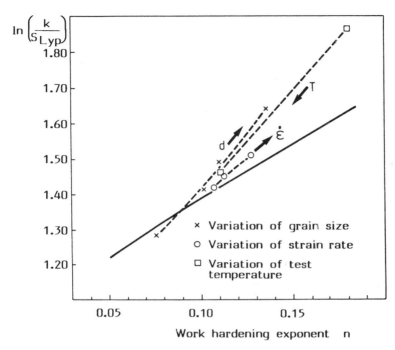

Figure 12. The solid line represents the stability criterion, eq.7. Materials showing combinations of parameters above the solid line undergo stable deformation. Arrows indicate increasing values of the parameters.

intersection between the curve describing the instability criterion and the extrapolated experimental curves defines the critical n-values and hence the maximum Lüders strain under the given conditions. It follows from fig. 12 that the deformation is becoming increasingly more stable as the work hardening exponent increases. Such an increase is observed with increasing grain size, increasing strain rate or decreasing test temperature.

The microstructural differences accompanying these changes are especially clear when the temperature is varied, since this parameter has a profound influence on the work hardening rate. Fig. 13a shows the initial grain structure in the material with grain size approximately 1.1μm. The structure is well developed with few matrix dislocations. Deformation at room temperature is accompanied by a low work hardening rate (n = 0.11). The microstructure in the deformed region close to the final fracture is shown in fig. 13b. It can be seen that the original grain structure is largely intact with a low density of matrix dislocations. Very few subgrain boundaries are created, indicating that the mean free path for the dislocations are of a magnitude comparable to the original grain size. Thus the original grain boundaries act as effective sinks for dislocations, causing a high rate of dynamic recovery and a low work hardening rate. At low temperatures, fig. 13c, the rate of work hardening increases. The undeformed structure can still be recognized after deformation at 77K but a number of subgrain boundaries have been formed and a high density of matrix dislocations are created. At high deformation temperatures, T = 573K, the dynamic recovery rate is high and the resulting structure, fig. 13d, indicate that work hardening is negligible.

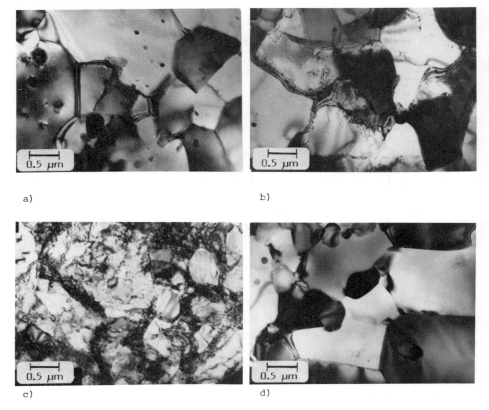

a)   b)

c)   d)

Fig. 13. TEM-micrographs showing: a) Undeformed grain structure,
grain size 1.1μm. b), c) and d) deformed to fracture at room
temperature, 77K and 573K, respectively.

The grain size has a similar effect on the deformation structures. At
sufficiently small grain sizes, the mean free path length for dislocations
becomes similar to the grain size and work hardening is reduced by efficient
annihilation of the dislocations. As the initial grain size increases, a
cell structure is built up in the interior of the grains, enhancing the work
hardening. This has also been demonstrated by Lloyd for a fine-grained
Al-6wt%Ni alloy (15).

The effect of strain rate on the microstructures is less pronounced
than the temperature and grain size effects. However, there is a tendency
that at the higher strain rates more subgrain boundaries and matrix dislocations are accumulated, reflecting the decreasing tendency for dynamic recovery at higher strain rates.

In ultrafine grained steels, a low test temperature or a high strain
rate favours the instabilities (1-4). For these materials the work hardening
rate is found to decrease with decreasing test temperature or increasing
strain rate, indicating that dynamic recovery effects are less important
than other factors which affect work hardening. The stability criterion, eq.
7, is valid also in the case of steel (1).

<u>Rate sensitivity of the flow stress and the work hardening rate.</u> The data in figs. 10 and 11 are presented in the form of Haasen-plots. It follows from thermal activation analysis that:

$$\left(\frac{\Delta\sigma}{\Delta\ln\dot{\epsilon}}\right)_{rev} = kT/b^3 \cdot \frac{b^2}{\Delta a} \qquad \text{eq. 11}$$

where k is Bolzmanns constant, T the absolute temperature, b the burgers vector and Δa the operational activation area (16). Thus the plot of $(\Delta\sigma/\Delta\ln\dot{\epsilon})_{rev}$ vs.σ gives the variation of the activation area with flow stress. Furthermore, since the work done during the thermally activated event, W is given by:

$$W = \sigma b \Delta a \qquad \text{eq. 12}$$

the slope of the $b^2/\Delta a$ vs. σ plot is given by b^3/W.

The variation of $b^2/\Delta a$ with flow stress at room temperature is linear at stress levels below about 150MPa, fig. 10. This is expected when the accumulated dislocations constitute a set of obstacles to flow. It can be noted that extrapolation to zero stress gives a positive intercept with the stress axis, indicating the presence of obstacles to flow which are more rate sensitive than dislocations (17). These obstacles are probably Mn atoms in solid solution. A prominent feature of the $b^2/\Delta a$ vs. σ curve is the strong departure from linearity at higher stress levels. The upward curvature has been observed by other workers (18, 19) and Mecking and Kocks have given a phenomenological explanation for the phenomenon (19). Their conclusion is that dynamic recovery is the direct reason for the nonlinearity. Both the flow stress and the work hardening rate are then temperature and strain rate sensitive, explaining the stabilization of flow as observed in fig. 6. Further evidence for the strong influence of recovery is provided by the increasing difference between $(\Delta\sigma/\Delta\ln\dot{\epsilon})_{cont}$ and $(\Delta\sigma/\Delta\ln\dot{\epsilon})_{rev}$ at higher flow stress levels. This difference is due to differences in the work hardening rates at the two strain rates, which again is caused by dynamic recovery effects. The behaviour at 77K is similar to the room temperature behaviour, however, the stress level where dynamic recovery effects become important are shifted to higher values, in agreement with the predictions of Mecking and Kocks (19).

As discussed by Lloyd (15), dynamic recovery becomes increasingly more important as the grain size decreases. We conclude that dynamic recovery and the associated decrease in work hardening capacity enhance the tendency for instability. This is further supported by the observed influence of strain rate and test temperature.

Summary

Instabilities during tensile deformation of fine-grained Al-alloys are initiated by an abrupt increase in mobile dislocation density, causing a drop in the flow stress and formation of Lüders bands.

The instability can be suppressed if the specimens are predeformed slightly by rolling, to activate dislocation sources throughout the gauge length.

Propagation of the Lüders bands depends on the rate of work hardening during Lüders deformation. A high work hardening rate tends to force the Lüders front along the gauge length. Below a critical work hardening rate, propagation of the Lüders band is suppressed, and the specimen fails under a decreasing load after the abrupt yielding.

The tendency for instability is enhanced by a decreasing grain size, an increasing test temperature and a decreasing strain rate. These factors affect the rate of dynamic recovery and hence the work hardening rate.

The described instability is of considerable practical interest. The onset of instability at small grain sizes limits the amount of useful grain boundary hardening in ultrafine grained Al-alloys. Furthermore, the localization of deformation and the accompanying premature failure of commercial purity temperannealed Al-sheet, limit the amount of subgrain hardening in such materials.

Acknowledgements

The author is grateful to the Department of Materials Engineering, Monash University as well as to Årdal og Sunndal Verk a.s. for permission to publish this work. He would like to thank Ian Dover for helpful discussions. Material was kindly provided by Norsk Hydro A/S through a project supported by Norges Teknisk-Naturvitenskaplige Forskningsråd (no. 7805.10324).

References

1. W.B.Morrison and R.L.Miller, in "Ultrafine - Grain Metals", ed. J.J.Burke and V.Weiss, Syracuse University Press, Syracuse, 1970, pp. 183-211.
2. W.B.Morrison, in Proceedings of the Second International Conference on the Strength of Metals and Alloys, American Society for Metals, Metals Park, 1970, pp. 879-883.
3. V.Ramachandran and E.P.Abrahamson, II, Scripta Met. $\underline{6}$, 1972, p. 287.
4. E.P.Abrahamson, II and V.Ramachandran, in Proceedings of the Fourth Bolton Landing Conference, ed. J.L.Walter, J.H.Westbrook and D.A.-Woodford, Claitor's Publishing Div, Baton.
5. L.R.Morris, in Proceedings of the Fourth International Conference on the Strength of Metals and Alloys, vol. 2, Nancy, 1976, pp. 649-653.
6. L.R.Morris, in Proceedings of the Conference on Solidification and Casting of Metals, Metals Society, Sheffield (1977), pp. 218-224.
7. I.Dover and H.Westengen, Aluminium, $\underline{60,}$ 1984, p. 741.
8. R.Iricibar, C.Pampillo and H.Chia, in Proc. Aluminium Transformation Technology and Applications, American Society for Metals, Metals Park 1980, pp. 241-303.
9. H.Westengen, in Proc. 6th International Conference on the Strength of Metals and Alloys, Pergamon Press, 1982, pp. 461-466.
10. E.Nes, J.Met.Sci., $\underline{13,}$ 1978, p. 2052.
11. H.Herø, in Proc. 10th Biannial Congress of the Industrial Deep Drawing Research Group, Warwick, 1978, p. 179.
12. A.H.Cotrell and R.J.Stokes, Proc. Royal Soc., A233, 1955, p. 17.
13. W.A.Backofen, in Deformation Processing, Addison-Wesley, 1972.
14. I.L.Dillamore, J.G.Roberts and A.C.Bush, Metal Sci., $\underline{13}$, 1979, p. 73.
15. D.J.Lloyd, Met.Sci., $\underline{14}$, 1980, p. 193.
16. U.F.Kocks, A.S.Argon and M.F.Ashby, Prog. Mater. Sci., $\underline{19}$, (1975).
17. R.A.Mulford, Acta Met., $\underline{27}$, 1979, p. 1115.
18. F.A.Bullen and M.M.Hutchison, Phil. Mag., $\underline{7}$, 1962, p. 557.
19. H.Mecking and U.F.Kocks, Acta Met., $\underline{29}$, 1981, p. 1865.

ON THE DEVELOPMENT OF STRUCTURE DURING THE EXTRUSION PROCESS

T. Sheppard, S.J. Paterson, M.G. Tutcher,

Department of Metallurgy and Materials Science,
Imperial College of Science and Technology,
London, S.W.7.

Summary

The structural changes occurring throughout the total thermomechanical extrusion process have been investigated and their effect on processing and subsequent properties are reported. It is shown that the product cycle is an initial step and adjustments to this cycle can result in improved processing conditions. The development of structure during the process is reported and macroscopic investigations indicate that existing mathematical models may not be sufficient to describe events in the deformation zone. It is shown that the development of a substructure commences in the die entry region and spreads from that location as deformation proceeds. Extrusion commences before the breakthrough pressure is achieved and the peak pressure is a necessary requirement to establish the quasi-static deformation zone. The variation of substructure in the product is reported and variations between direct and indirect extrusion are discussed. It is shown that a substructure can be retained after the solution soak and ageing treatment and the effect on mechanical properties of alloy 2014 are discussed.

Introduction

The extrusion process is most often utilised to produce an engineering product which must satisfy strict geometric, cosmetic and property specifications. A proportion of material is also extruded as forge stock or requires further machining. The process is complex, involving interaction between the material high temperature properties and the extrusion variables. The material properties presented to the extrusion press may be modified from the cast structure by an homogenisation treatment and by the necessary reheat prior to working. During the working process the structure is further modified and is determined by the temperature, ram speed and extrusion ratio obtaining. These process parameters may be incorporated to calculate a temperature compensated strain rate. The relationship between the subsequent structure and the temperature compensated strain rate is now well established, having been determined during creep (1), high temperature testing (2), (3), and extrusion (4) - (7). The work reported in this communication has been conducted at Imperial College over a period of years and concerns the extrusion of a number of Aluminium alloys.

Experimental Procedure

The experimental details have in general been described elsewhere (8) - (11), but reference to relevant details is made in the text. The materials were supplied by ALCAN INTERNATIONAL LTD. in the form of 85mm direct chill cast logs which were subsequently turned and sawn to produce 75mm dia. x 120mm long billets. The chemical composition of the alloys is given in Table 1.

Table 1.   Composition of Alloys, wt%

Designation	Cu	Si	Mn	Mg	Fe	Zn	Ti	B	Al
AA 1100	0.002	0.001	–	0.001	0.22	0.01	0.012	–	Bal
AA 2014	4.53	0.82	0.81	0.70	0.17	0.02	0.013	0.0014	Bal
Al-5Mg	–	0.003	0.001	4.9	0.001	–	0.001	–	Bal

Results and Discussion

Structural Changes During Homogenization (AA 2014 Alloy)

Fig.1 shows optical micrographs of the billet structure after soaking at 500°C for 8, 16 and 24h, respectively. In each case the billet was air cooled. There is considerable modification to structure after 8h soaking, but the dendrite arms are still clearly visible within the recrystallised grains. Further soaking results in the diffusion removal of most of the cast structure (Fig.1a and c) but even after 24h some evidence of the dendritic structure still remains. It is also clear that, as expected, there is no substantial grain growth during homogenisation. The structures were also examined at higher magnification in the scanning mode of the TEMSCAN. Fig.2a shows the structure after exposure at 500°C for 8h: segregation has been very considerably reduced and the soluble components redistributed throughout the structure by diffusion. That part of the original structure which has been retained has left a

FIGURE 1(a)

FIGURE 1(c)

FIGURE 1. Billet structure after homogenisation at 500°C and air cooling.
(a) 8h; (b) 16h; (c) 24h.

FIGURE 1(b)

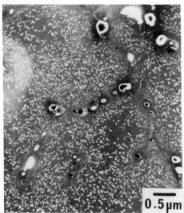

FIGURE 2(a)

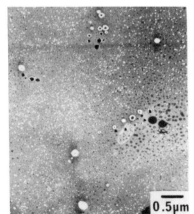

FIGURE 2(b)

FIGURE 2. Billet structure after homogenisation
(a) for 8h at 500°C
(b) for 24h at 800°C.

residue of large particles in the intergranular regions. The effect of soaking at 500°C for 24h is shown at higher magnification in Fig.2b. Comparison with Fig.2a demonstrates the advantages of prolonged exposure at high temperature. The structure consists of a uniform distribution of small precipitaes with only a few isolated large particles. One of these large particles was examined in transmission mode and was identified as a Mg_2Si particle in a precipitate-free area. This would suggest that the increase in copper taken into solution during homogenisation forces the silicon to combine preferentially with the magnesium to form Mg_2Si. The size of the Mg_2Si particles was typically 0.6µm, compared to the 1.5µm particles found in the material soaked for only 8h. Thus the extended soak enabled further diffusion to reduce the size of the larger particles. The smaller particles evident in the structure were identified as $(Fe.Mn)Al_6$ and $MnAl_6$, having a size of about 0.05 - 0.075µm).

It is thus obvious that homogenisation time and temperature are important variables in the thermomechanical process. For this particular alloy it would appear to be desirable to utilize an extended soak at 500°C. The same metallurgical result could, however, presumably also be achieved at some other time in the cycle. Hence it will be instructive to investigate the effect of homogenisation on the extrusion process.

Effect of Homogenisation on Extrusion Variables (AA 2014) Alloy)

The preceding section has indicated that the time and temperature involved in the homogenisation cycle have a pronounced effect on the structure of the material which enters the extrusion container. The effect of these structures on the extrusion pressure and the resultant structure of the extrudate is now considered. Billets were extruded in the as-cast condition, after homogenisation for 24h at 450 C and after homogenisation for 24h at 500°C.

The effect of the initial billet structure on extrusion pressure is shown in Fig.3: it is evident that the initial structure quite drastically affects the extrudability of alloy 2014. The as-cast material requires about 20% more pressure than that required to extrude a billet given a thorough homogenisation treatment. The result of gross segregation is that hard brittle eutectics and large volume fractions of precipitate are present during the extrusion process: these raise the flow stress of the material, consequently raising the required extrusion pressure. The billets homogenised for 24h at 450°C required about 10% additional pressure, confirming that the continued presence of partial segregation is deleterious. The excess pressure also implies that the temperature rise during the extrusion process will be greater in the as-cast material, leading to more difficulty in controlling surface defects. The extrusion limits are thus generally reduced when extruding incorrectly homogenised material.

The structure resulting from the extrusion of an as-cast billet is shown in Fig.4, the dominant feature of the morphology being the stringers of large particles aligned in the extrusion direction. These particles have the as-cast eutectic as their origin and have

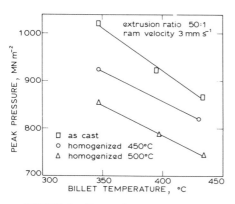

FIGURE 3. Extrusion-pressure
dependence on initial billet
temperature.

FIGURE 4. Extrudate structure of
as-cast material.

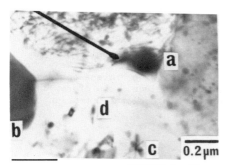

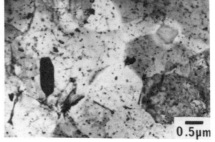

FIGURE 5. Microstructure of 440°C
extrudate, fully homogenized
(500°C, 24h) before extrusion.

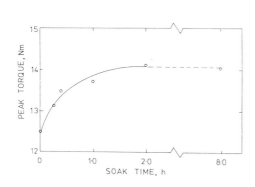

FIGURE 6. Variation of Peak
Torque with soak time
$T_i = 500°C$, $T_e = 425°C$, $\dot{\varepsilon} = 8.5 \text{ s}^{-1}$

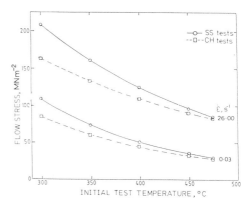

FIGURE 7. Comparison of flow stress
for conventionally preheated (CH)
and solution soak (SS) specimens.

127

been fragmented by shear during the extrusion process. EPMA analysis identified them as the Mg_2Si and $CuAl_2$ phases. Thus the segregation observed in the as-cast structure is retained through the extrusion process, becoming detrimental to the normal age-hardening process. It might also be expected that these stringers and large particles would be detrimental to the extrudate fracture properties. Fig.4 also demonstrates that the substructure is affected by these stringers taking on a 'brick-like' morphology, rather than the usual, more equiaxed subgrain structure. The structure of the extrudates produced with an extended homogenisation treatment (24h at 500°C) show a reduction in the number of large particles and a more uniform distribution of small precipitates. The structure shown in Fig.5 was analysed, the large particles being identified as Mg_2Si and the smaller particles as $(FeMn)Al_6$. The substructure in all the extrudates produced from these billets was more equiaxed, with a complete absence of stringers and generally a finer morphology, than the structures observed in the other extrudates. A significant feature revealed by the analysis was the notable absence of copper-bearing particles, indicating that most of the copper was in solution.

The results, therefore, indicate that prolonged homogenisation is necessary to avoid non-uniform distributions of large copper-bearing precipitates, which will certainly be detrimental to the tensile and fracture properties of the F-temper materials. High extrusion temperatures or subsequent solution treatment could modify the copper distribution in the unhomogenised cases, but the quantity and distribution of the large particles is likely to persist: thus the adverse effect on the extrusion process could not possibly be avoided. It is possible to conclude, therefore, that an extended homogenisation is essential before the extrusion of Al-alloys.

Preheat Variation (AA 2014 Alloy).

Torsion Experiments

To establish whether heating prior to extrusion could affect extrudability, torsion tests were conducted aimed at comparing the constitutive equation for varying structures. They were performed using a statistical central composite matrix performed on the material which had been extruded and subjected to a solution soak for 24h at 500°C and furnace cooled. The results were fitted to the equation :

$$\sigma = \frac{1}{\alpha} \ln \left[(Z/A)^{2/n} + < (Z/A)^{2/n} + 1 >^{\frac{1}{2}} \right] \qquad (1)$$

giving a correlation coefficient of 0.985 and material constants $\Delta H = 14408$ J/mole, $\alpha = 0.0152$ N^{-1} .mm^2 , $n = 5.27$, $A = 2.957$ x 10^{10} s^{-1} ,$\alpha n = 80.104$ x 10^{-3} . In this equation σ is the flow stress, α is a constant when the peak torque is used for evaluation, Z is the Zener Hollomon parameter = $\dot{\epsilon}$ $\exp{\frac{\Delta H}{GT}}$, n is the strain rate sensitivity, ΔH the activation energy for deformation, G the universal gas constant, and A is a material constant.

The succeeding series of experiments evaluated the minimum soak time required to produce the maximum flow stress. The specimens

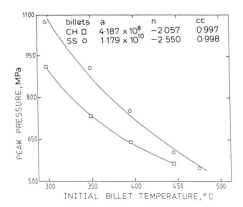

FIGURE 8(a). Effect of modified
preheat on extrusion breakthrough
pressure for direct extrusion.
R = 20:1, v = 13 mm s^{-1}.

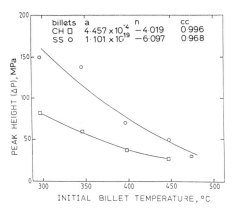

FIGURE 8(b). Effect of modified
preheat on pressure increment
required to initiate extrusion.
R = 20:1, v = 13 mm s^{-1}.

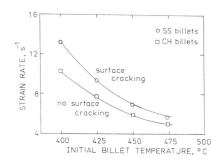

FIGURE 9. Strain Rate required for
surface cracking v T_i for direct
extrusion.

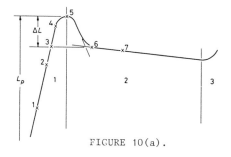

FIGURE 10(a).

FIGURE 10(a). Load/displacement
curve showing peak geometry
and sample locations.

FIGURE 10(b). Positions of
electron microscopy specimens.

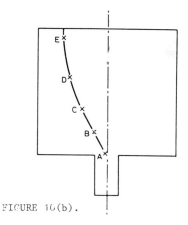

FIGURE 10(b).

129

were heated to $500^\circ C$ for varying times from 0 to 8 hours and allowed to air cool to a typical extrusion temperature of $425^\circ C$. They were then soaked for 2 mins to attain a uniform temperature and tested as described previously. The results are shown in Fig.6 which indicates that a minimum soak time of two hours is required to obtain the maximum increase in flow stress due to solid solution strengthening. Consequently a torsion matrix was designed to evaluate the flow stress in which the specimens were soaked for two hours and cooled in the induction coil to temperatures ranging between $275^\circ C$ and $450^\circ C$. The constants derived using the minimisation technique were

$$\Delta H = 176867 \text{ J/Mole}, \ \alpha = 0.01181 \text{ N}^{-1} \text{ mm}^2, \ n = 5.86, \ A = 4.466 \times 10^{13} \text{s}^{-1},$$
$$\alpha n = 692066 \times 10^{-3}$$

with a correlation coefficient of 0.984. A comparison of the flow stress to that obtained without the solution soak for high and low strain rates indicated that the flow stress was always higher in the material subjected to the heat treatment designed to increase the copper retained in solution, an effect retained even at the lowest test temperatures.

The value of the function (αn) may be regarded as representing the activation volume or the magnitude of the obstacles opposing dislocation motion. Thus a decrease in the value of this function should result in an increase in stress since we may also regard it as a measure of the effectiveness with which the applied stress aids dislocations in overcoming barriers. The results indicate that this factor decreased by the ratio 1.15 after the prolonged soak which is consistent with the above interpretation. The results also indicate an increase in the value of n or decrease in strain rate sensitivity which should improve the reaction to ram speed during the extrusion process. A stationary dislocation in a solid solution can be pinned by the solute atoms if they are able to diffuse to the dislocation and hence an addition increment of stress will be required to free the dislocation from its energetically favourable position. In addition, any dislocation moving through a solid solution will encounter friction drag thus raising the energy required for movement. Hence any increase in foreign atoms held in solution will increase the activation energy, ΔH. In this case we might expect an increase in the Cu, Mg and Si held in solution by the preheating sequence which explains the higher measured activation energy. During high strain rate deformation, recovery processes are also hindered by a higher solute atom vacancy binding energy (12) which effectively reduces the number of vacancies available for dislocation climb whilst the misfit strain effectively raises the dislocation density; both processes contributing to an increase in activation energy. The material characteristic A in the constitutive equation is a measure of the frequency with which a dislocation may overcome an obstacle. It is not completely independent because with any increase in activation energy the constant A must increase to avoid excessive increases in flow stress; there are many more obstacles to overcome hence the frequency must increase as indeed it does for this particular material configuration.

Based upon these results an experimental extrusion matrix was designed and tested using a solution soak of 2h incorporated into the preheat cycle. Fig.7 shows the effect of the revised treatment on the breakthrough pressure required for the extrusion process. The results are as expected, indicating that greater pressures are required consistent with the increased flow stress of the material. The slower cooling rate in the extrusion cycle does not seem to have a significant effect indicating the slow precipitation of the transition phase in the 2014 alloys. It is also interesting to note that the increased pressure is almost entirely due to the increment of pressure required to effect breakthrough Δp, as shown in Fig.8. This is consistent with the theory that this phenomenon is associated with the requirement of generation of excess dislocations in order to establish the extrusion deformation zone. In this case it would appear that the additional solute content impedes dislocation motion to such an extent that it is only this initial pressure (rather than the running pressure) which is greatly affected. This is significant because it affects the total work done on the billet and hence the energy transformed into heat; the reader should recall that it is towards the end of the extrusion stroke that surface quality and properties are likely to become unacceptable.

The surface quality of the extrudates was assessed by visual inspection and categorised as (A) acceptable surface finish, (B) showing evidence of incipient melting or (C) showing excessive die lines or surface tearing. Analysis of the results indicated that the line separating acceptable surface finish (A) and poor or unacceptable surfaces (B,C) was clearly defined and it is apparent that the billets incorporating a solution soak in the preheat offer considerably greater resistance to surface defects. The defining lines were fitted to power law relationships such that the achievement of an acceptable surface finish was defined by :

$\ln Z_i < 68000$ T   for conventional heated billets

$\ln Z_i < 98000$ T   for the pre-solution soak billets

The differences defined by these equations are, in part, due to the large difference in activation energy between the two materials. However, when the strain rate at which cracking occurs is evaluated against the extrusion temperature (Fig.9), it is clear that substantial increases in ram speed are possible when utilising the two step preheat. For example, at an extrusion temperature of 400 C the maximum ram speed may be increased from approximately 12.6mm/sec. at an extrusion ratio of 20:1.

Development of Structure (Experimental Alloy Al-5Mg)

To investigate the mechanisms involved in the development of substructure billets were partially extruded, removed from the container and rapidly quenched. The arresting locations on the load displacement loci and the positions from which foils were extracted for examination by electron microscopy are shown in Fig.10. A recrystallisation inhibitor was purposely omitted from the melt in order that the macrostructure would be used to define the flow characteristics. The grains were consequently relatively large, especially towards the periphery of the billet. There was no significant volume fraction of precipitate. During the upsetting stage (location 1), the macrostructure did not vary from that of the original billet with the exception of the region adjacent to the die entry where some deformation could be detected. There was no evidence of deformation elsewhere in the billet although container/billet shear is suggested by the grain structure at the periphery. Fig.11a, illustrating the structure at a later stage in the upsetting process, does not exhibit radical changes, but an extension of the main deformation zone has occurred and the grains in this die entry region have become elongated. It is interesting to note that a small quantity of material has been forced through the orifice and that the extreme front end reveals quite clearly the original undeformed structure. Even at this stage, however, there is a variation in structure from front to rear of the extruded pip. Fig.11b shows the situation after further increase in pressure but with little ram travel and shows the deformation zone extending back into the billet. There is still little change in bulk structure and no identifiable dead-metal zone but cylinder wall shear appears to have extended over the complete billet surface. Fig.11c corresponding to location 4 in Fig.10 shows an almost identical macrostructure but the fibrous nature of the extruded structure suggests an increase in total deformation. The structure in Fig.11d located approximately at the point of peak pressure shows very pronounced changes in morphology. The extrudate exhibits the typical fibrous or 'cold worked' structure and it should be noted that at this stage a considerable length of material has been extruded. Clearly the 'breakthrough' pressure does not correspond to the commencement of extrusion. The deformation zone has again extended backwards into the billet and dead-metal zones are becoming apparent. Considerable cylinder wall shear is apparent and the billet peripheral layers are quite heavily deformed. The latter feature is less predominant in the surfaces adjacent to the die plane, supporting the observation that the dead-metal zone will be established in this area. In fact the deformation zone extends well back into the billet and grains can quite easily be identified which have been elongated within the central deformation zone. Fig.11e shows the structure when the pressure has fallen by an amount of ΔL after the peak pressure and considerable ram travel has occurred. The central deformation zone has now developed fully and intense shear is occurring at this location; cylinder wall shear also appears to be more intense, but it should be recalled that it is a deformation history which is being observed since quasi-static conditions do not obtain at this location. It should also be noted that although the dead-metal zones can be identified clearly it is the lessening of surface shear and gradation of grain deformation which allows this observation; there is definite evidence that the quasi-static deformation zone is about to be established. Fig.11f illustrates the macrostructure in the steady state region. The quasi-static deformation zone is fully developed and the dead-metal zone can very easily be identified. Regions of significant shear can be seen extending from die entry and along the central axis and from die entry to the periphery of the billet bounded by the dead-metal zone and the container/billet interface. The large initial grain size allows the recognition of grains passing through three regions of intense deformation. The identification of two regions of lesser deformation on either side of the central zone and sandwiched by it and the 'peripheral' zone suggests a buffer zone which serves primarily to supply material to the regions of high deformation in the billet. Material flowing from the peripheral

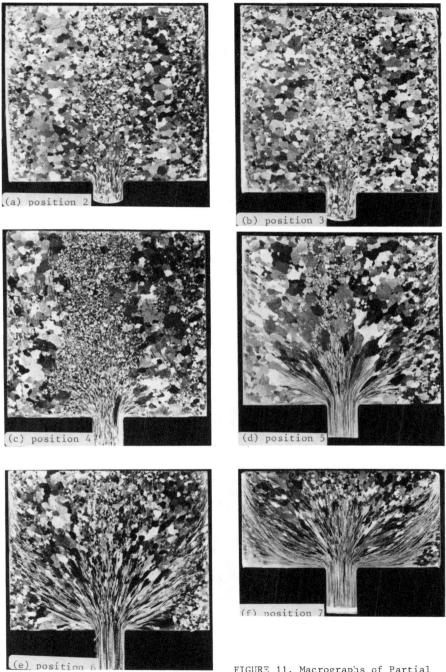

(a) position 2

(b) position 3

(c) position 4

(d) position 5

(e) position 6

(f) position 7

FIGURE 11. Macrographs of Partial extruded billet.

region of the billet can be seen to occupy about 35% of the extrudate cross-section while material from the axial deformation zone occupies about 55%. This implies that the two zones of lesser deformation are feeding only about 10% in total of the material section shown in Fig.11f. Clearly, if these zones were not primarily feeding the intense shear zones, vortices would form within the deformation zone.

In summary, it appears that significant macrostructural changes occur only after the peak pressure. During upsetting the deformation zone extends backward from the die mouth and along the cylinder wall and dead metal zone. At the peak pressure, deformation is well developed but steady state macroconditions develop between the instant of peak pressure and a time corresponding to a decrease of ΔL from this point. The steady state region is characterised by much greater intensity of shear at the axis and periphery of the billet.

Development of Substructure (Experimental Alloy Al-5Mg)

The dislocation structure within the billet arrested at position 1 in Fig.10a was heterogeneous. Specimens extracted from the rear of the billet (positions C, D and E) had similar structures having a uniform dislocation density. In isolated areas, however, walls of dislocations were beginning to form. Similar walls were observed in the other rear positions indicating that the upsetting operation causes deformation throughout the billet. There did not appear to be any special deformation associated with possible ram/billet interaction. Further forward, in position B, dislocation wall formation was more pronounced and outcrops of subgrains were observed at high-angle boundaries. In general subgrains were not observed consistently at any position in this billet, which is consistent with the thesis that strains imposed on the material during the upsetting process are absorbed by shear at the grain boundaries. This will involve the production of dislocations to produce the strain and additional dislocation generation for storage within grains to maintain the strain gradients imposed by grain boundary shear. Thus the dislocation density is necessarily increased and a driving force for subgrain formation created. Fig.12a illustrates this feature. Although most grains were devoid of subgrains, the specimen extracted from the die mouth (position A) showed a subgrain banding phenomenon. Bands of subgrains were observed bisecting the grains and being elongated quite severely in the potential extrusion direction (at this stage extrusion has not commenced). Misorientations were higher between these bands than between the subgrains along them, which can be seen in Fig.12b by noting the small contrast between the individual subgrains and the large difference in contrast between these and the remainder of the grain. This was confirmed by measurement of misorientations from selected-area diffraction patterns. The banding would seem to be the result of strains exerted upon abnormally large grains as they pass the velocity discontinuity associated with entry to the die mouth. These grains are preferentially elongated and the resulting localized regions of high strain require that areas of higher dislocation density also exist. Hence in this area there is sufficient dislocation mobility and density to effect the formation of subgrains. This observation may be even more significant in the industrial context where we would expect greater heterogeneity in the 'homogenized' billet by larger grain sizes, more diverse dendritic morphology, and the presence of dissociated soluble constituents.

Specimens extracted at locations 2 and 3 during the process exhibited similar structures. Banding was once more observed in the die entry region: in this case producing more equiaxed subgrains as shown in Fig.12c. It should also be noted that in these subgrains there is considerable internal dislocation density and, in general, the walls are more clearly defined than those observed earlier in the extrusion cycle. The selected-area diffraction patterns (Fig.12c) show that the misorientation between the bands consists of a lattice rotation of 33° which reverses in sign across alternate bands. This was generally the case for all such bands investigated. At locations towards the rear of the billet dislocation wall build-up is more pronounced and isolated outcrops of small subgrains emanating from high-angle boundaries can be identified. Dislocation wall formation was also observed to be aided by precipitates; these sites clearly act as locking points and extended walls can be observed which have their extremities at precipitates. Humphreys (13) has shown recently that very large lattice rotations may occur at second-phase particles and this will require high dislocation densities. The probability that subgrain wall formation will develop is thus enhanced at these sites. The dislocation density within subgrains indicates incomplete recovery, and within completely unrecovered regions the dislocation density is much greater than earlier in the ram stroke.

Specimens taken from the billet corresponding to the point of peak pressure (location 5) show an equiaxed subgrain structure in positions A, B and C. These subgrains are slightly elongated in the die mouth (Fig.12d) and are heterogeneous in size, being smaller at high-angle boundaries. It is interesting to note that the dislocation density within subgrains is lower in specimens B and C (i.e. proceeding backwards into the deformation zone) which is consistent with lower strains and lower strain gradients in these regions than those which occur around the theoretical exit-velocity discontinuity. Nevertheless it is clear that at this stage the deformation zone is not quasi-static and the dead-metal zone has not been fully developed since areas can be identified within the deformation zone in which subgrain formation is incomplete and the macrographs indicate that the dead-metal zone is not well defined.

When the ram has advanced to location 6, an equiaxed subgrain structure can be observed in positions A, B, C and D. Some banding can still be detected at position A (Fig.12c) but, in general, subgrains are equiaxed, locally smaller at high-angle grain boundaries and have a high internal dislocation density (Fig.12f). The specimen from the back of the billet contains only small outcrops of subgrains located at high-angle grain boundaries and intergranular dislocation walls. These subgrains and dislocation wall formations are most probably remnants from compression during the upsetting stage. Although this deformation is not detectable in the macrosections it is clear that compressive flow must occur throughout the billet during the initial ram contact. Proof that large strain gradients are associated with high-angle grain bondaries is shown in Fig.12g. This depicts a section of high-angle grain boundary involving a bend through 90°. Subgrains which are formed heterogeneously at high-angle grain boundaries can be seen to be much better defined along the boundary aligned parallel to the billet axis (and the

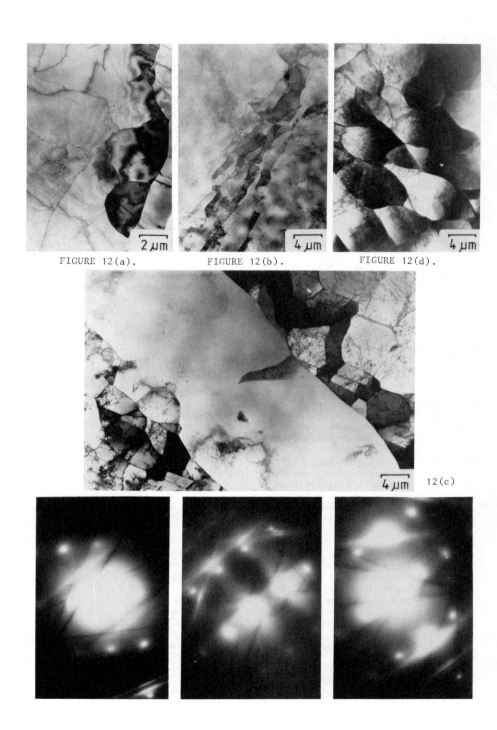

FIGURE 12(a).　　　FIGURE 12(b).　　　FIGURE 12(d).

12(c)

FIGURE 12(e).   FIGURE 12(f).

FIGURE 12(h).

FIGURE 12(g).

FIGURE 12(a). Subgrain formation at high angle grain boundary.

FIGURE 12(b). Bands of elongated subgrains formed in material at die mouth (position A) during initial upsetting of billet (location 1 in load/extension curve).

FIGURE 12(c). Deformation bands containing subgrains formed in large grain. Sample taken from position A in billet at location 2 on load/displacement curve.

FIGURE 12(d). Elongated subgrains containing very high dislocation density observed at die mouth at peak pressure. Dynamic recovery is incomplete.

FIGURE 12(e). Equiaxed subgrains within deformation bands in sample taken from the same position as that in Fig.12(d). but at location 6 in the load/displacement curve.

FIGURE 12(f). Sample containing high-angle grain boundary (arrowed) taken from the same location and position in the load/displacement curve as Fig.12(e).

FIGURE 12(g). Specimen taken from the back of the billet at location 6 in the load/displacement curve.

FIGURE 12(h). Typical subgrain and dislocation structure observed during steady state extrusion. Sample taken from position B in billet at location 7 in load/dispacement curve.

137

vertical axis of the micrograph) and at the grain extremity than along
the segment at right angles to it. The latter is a region of much
smaller strain gradient. Tilting experiments showed that this is a
genuine effect and not due simply to a particular set of imaging
conditions being employed.

Samples inspected from the steady-state position (location 7)
revealed equiaxed subgrain structures. In locations A and B the
subgrains were of uniform size but location C included smaller
subgrains at high-angle boundaries. Presumably this phenomenon is
associated with a less acute breakdown of original structure than has
occurred at later stages in the deformation zone, i.e. locations A and
B. Original high-angle grain boundaries were impossible to find in
the deformation zone exit region because of the intense deformation
history of the material in that location. Typical structures are
illustrated in Fig.12h. The dislocation density within subgrains is
higher at the deformation zone exit than at position C which is
consistent with the very rapid increase in strain rate (and hence
strain) which occurs in that region (14).

Comparison of Direct and Indirect Extrusion (AA 1100 Alloy)

The development of flow during the indirect extrusion process is
illustrated in Fig.13 which shows macrographs of partially extruded
billets at three stages during the extrusion cycle and indicates the
location of electron microscopy sampling. The tongue visible at the
rear of the billet is formed when material flows back into the dummy
block, which is so shaped to facilitate discard removal. Fig.13a
depicts the flow pattern at peak pressure and again it is evident that
extrusion commences sometime before this because the resultant
extrudate quite clearly has a well-developed fibrous structure. Two
deformation zones are evident: one adjacent to the die entry and one
associated with the tongue. The latter will be ignored because it is
purely functional and does not change during the extrusion process.
The former illustrates that the metal flow is complex. There are
semi-elliptical dead-metal zones associated with the die face bordered
by shear zones which apparently have their origin at the billet
surface. These zones must however be fed, in part, by the material
immediately to the rear of the billet. However, it is clear that at
this stage the surface fibres of the product have their origin at the
billet surface and that upsetting has caused deformation which is much
heavier at the centre of the billet than elsewhere. The localised
regions of internal shear which can be observed at the extremities of
the billet/die interface are associated with localised flow and the
hollow stem and occur during the initial stages of extrusion as the
material attempts to back-extrude through the container/stem
clearance. Fig.13b is a macrograph from a billet arrested during
steady-state extrusion when the quasi-static deformation zone is fully
established. It is noticeable that the shear bands are not as
intense as those observed during direct extrusion: indeed the regions
of heaviest shear appear to form the central regions of the extrudate.
Contrary to expectations, a dead metal zone can still be distinguished
at the die face, material adjacent to this zone being more heavily
sheared and forming the surface of the extrudate. Thus the
acceptable surface finish during indirect extrusion is produced by the
same shear mechanisms as those operating during direct extrusion but
these are of lesser intensity. The localised deformation at the

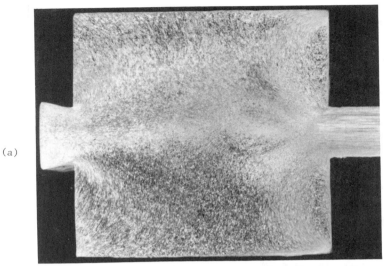

(a)

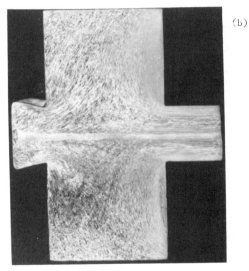

(b)

(c)

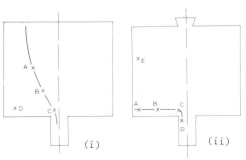

(i) (ii)

FIGURE 13.
(i) Location of electron
    microscopy samples in
    direct extrusion.
(ii) Location of electron
    microscopy samples in
    indirect extrusion.
(a) Macrostructure of
    indirectly extruded billet
    partially extruded to peak
    pressure.
(b) extruded to quasi-static
    state.
(c) showing rear end effect

139

extremities of the hollow stem has ceased and it is obvious that flow occurs from the less deformed regions to the heavily deformed regions as is the case for direct extrusion. Fig.13c shows metal flow close to termination of the extrusion cycle when further ram movement would result in material from the tongue region being forced through the die, and a significant rise in extrusion pressure. The macrograph depicts almost uniform deformation with material flowing along elliptical stream-lines from surface to centre of the billet. A specific feature of these sections is that, unlike direct extrusion, the structure of the extrudate appears to be virtually homogeneous throughout the cross-section. This latter comment is more or less true of all the micrographs in Fig.13 which suggests that if there is a region of heavier deformation in the extrudate then it is at the centre rather than at the surfaces.

Figs. 11 and 13 indicate that material flow is far more homogeneous in the indirect process resulting in a more uniform extrude. The surface layers in indirect extrusion originate in the surface of the billet. This factor will necessitate scalping of the billet prior to extrusion and probably means that only those alloys utilised in high technology areas should be processed by the indirect route (because such billets would be scalped prior to direct extrusion). The other interesting result of the more homogeneous flow is that the surface layers originate from material having a different strain - and therefore strain rate and temperature - history than in direct extrusion. This greatly decreases the incidence of surface cracking and recrystallisation and hence ensures that the indirect process is an attractive route.

In order to investigate the substructural development, extrusion was interrupted during the steady-state phase for both the direct and indirect processes. The extrusion in each case was performed at an extrusion ratio of 40:1 a ram speed of 7.5mm/s and an initial billet temperature of 300 C. Specimens were removed for examination from the locations shown in Fig.13[i] and 13[ii] selected such that the development of structure along the flow lines indicated could be analysed.

The development of structure along a flow line for the direct extrusion process is shown in Fig.14. Fig.14a is from a position which is just inside the rear extremity of the deformation zone (position A). A well-defined substructure has already developed with the subgrains exhibiting a high internal dislocation density. The subgrains are equiaxed and were observed to be heterogeneous in size. The role of the small precipitates, which are probably Fe compounds, in pinning internal dislocations can easily be observed. Figs.14b and 14c are micrographs corresponding to positions B and C in Fig.13a and indicate a progressive decrease in internal-dislocation density and in size. The decrease in size, determined by the intercept method, was found to be from 2.9 µm at position A to 2.5 µm at position C. A noticeable feature was that subgrain banding which had been observed in the more complex alloys was not present and indicates that such banding phenomena may well be connected with the dislocation mobility being modified by large solute contents and with second-phase particles. Fig.14d is from the dead metal zone and shows a substructure which is poorly developed but in which subgrains can clearly be identified. This deformation is clearly due to compression of the dead-metal zone but again differs from most

140

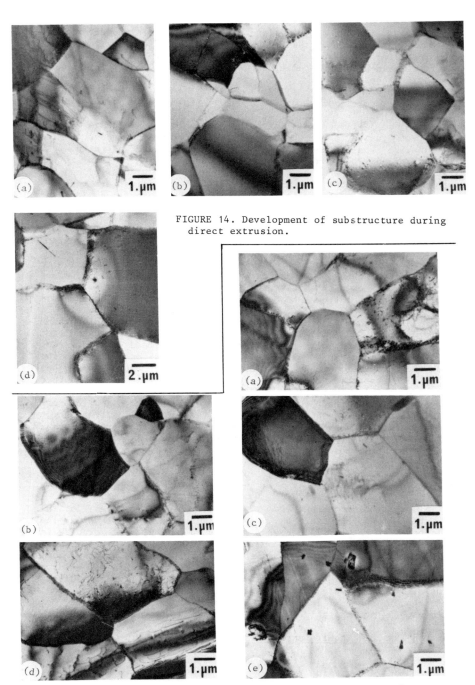

FIGURE 14. Development of substructure during direct extrusion.

FIGURE 15. Development of substructure during indirect extrusion.

commercial aluminium alloys which in general do not exhibit subgrain formation at this location. The size of the subgrains was observed to be 6.0μm considerably greater than that found in the deformation zone but indicating that the production, at the die entry, of small subgrains is progressive and that the subgrain size is related to the specific strain-rate and temperature conditions at each point in the deformation zone.

The development of structure along a flow line for extrusion in the indirect mode is shown in Fig.15. Micrograph (a) corresponds to position A in Fig.13d and is of a specimen close to the edge of the billet but inside the indirect extrusion deformation zone. The subgrains are well defined but some walls are ragged and there is a high internal-dislocation density. The corresponding positon in the direct billet would be the dead-metal zone. Micrographs (b) and (c) correspond to positions B and C in the billet and illustrate that the subgrains gradually become perfect and contain internal-dislocation structures only when precipitates are available to pin the structure. It is noticeable that the diminution in subgrain size is not as great as in the case of direct extrusion and in fact the subgrains are larger in these locations than at locations A, B and C in the direct case. This is a strong indication that the strain rate is much lower and since the average strain-rate is higher in indirect extrusion it is not surprising that the micrograph from position D (Fig.15d) shows a dramatic change with an average subgrain size of 2.08 m. Thus material traversing from A to D has the subgrain size transformed in the sequence 4.3, 4.3, 3.52 to 2.08 μm. It was not possible to extract a specimen from the dead-metal zone observed in the optical micrograph but micrographs (a) - (d) clearly indicate that the flow in indirect extrusion is quite different from that in steady-state direct extrusion. Micrograph (e) corresponding to positon E is located inside the deformation zone predicted by the upper-bound theory but just outside that observed in the optical micrograph. A substructure has formed but the subgrains are large (5.4μm), are not completely equiaxed, have rather ragged walls and have a high internal dislocation density . Clearly there is very limited deformation at this location. Micrograph (f) is of material extracted from the rear of the billet and shows a substructure comparable with that observed in the dead-metal zone in direct extrusion. The structure was most probably formed due to a combination of compression and the formation of the tongue during upsetting but generally indicates a lack of deformation in this area.

It can thus be concluded that there are differences in the development of substructure during direct and indirect extrusion. The development of the final subgrain size is a more gradual process in direct extrusion where the deformation zone is larger. It is clear that very large strain-rate gradients exist in the die entry zone during indirect extrusion. It is likely therefore that structures will be more consistent during direct extrusion than during indirect extrusion because localised heterogeneity will affect the final substructure in the indirect case where the subgrain size is determined in a comparatively short period of time.

Effect on Pressure Requirements (AA 2014)

The pressure required to extrude metal has been recognised to be dependent on the strain rate $\dot{\varepsilon}$, the extrusion ratio R, and the

temperature of extrusion T.   Mathematical models based on slip-line
field or upper-bound analysis can be used to predict that the pressure
will be given by an equation of the form

$$p = \sigma (x + y \ln R)\dots\dots\dots\dots\dots\dots [3]$$

in which p is the extrusion pressure, σ is the mean equivalent flow
stress, R is the extrusion ratio, x and y are extrusion constants.
This expression is usually derived by assuming that the material is
perfectly plastic.   Therefore, it does not recognise that there is an
additional increment of pressure required for 'breakthrough' of the
extrudate, which is alloy dependent.   The increment of pressure p
required to initiate extrusion was evaluated using a tangential
construction which is shown in Fig.16 together with the results
obtained.   This shows clearly that the additional increment is a
separate thermally activated event, and that for any given set of
process conditions the pressure required is lower for indirect than
for direct extrusion.   Thus, the indirect-extrusion route benefits
from improved friction conditions and from changes in dislocation
activity within the deformation zone.   The linear relationships
obtained from a regression analysis, which included results from all
the experiments, were :

$$p = 11.27 \ \ln Z_p - 250.9 \text{ for direct extrusion}$$

$$p = \ \ 9.8 \ \ln Z_p - 227.9 \ \text{ for indirect extrusion}$$

where Z_p is the temperature compensated strain rate at peak pressure
(the temperatures having been calculated using an integral profile
method).

    The ratio of p to total extrusion pressure requirements is
greater for direct than for indirect extrusion, suggesting that the
greater shear zones observed in direct extrusion and the differing
forms of deformation flow ensure that fewer excess dislocations are
required to form the deformation zone in indrect extrusion than in
direct extrusion.   The general reduction of pressure during the
extrusion process has always been attributed to the reduction in
friction as the billet length decreases. The pressure curves observed
for this alloy indicate that the reduction in pressure is so
non-linear that this explanation cannot be justified.

Summary of Substructure Development.

    The work reported here indicates that dislocation generation and
motion commences and spreads from the die entry as the pressure is
applied.   Dislocation density increases rapidly as a consequence of
the interaction of mobile dislocations with each other and with
dislocations stored within grains.   These latter dislocations are a
necessary consequence of the strain gradients which exist both on a
macroscopic scale (e.g. between the deformation zone and regions of
undeformed material) and on a microscopic scale i.e. across grain
boundaries at which multiple slip is required from the start of
deformation in order for the grains to remain conterminous.
Dislocation density increases most rapidly in these regions in which
strain and strain gradients are highest.   Grain boundaries, at which
dislocations are generated, are necessarily such sites.   Eventually

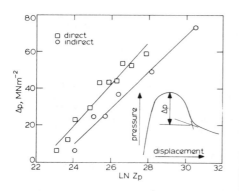

FIGURE 16. Construction of pressure evaluation and, inset, variation of peak increment v temperature-compensated strain rate.

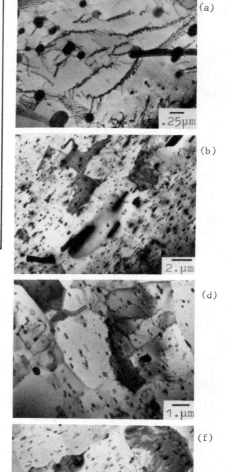

FIGURE 17. Substructure observed in the horizontal plane of the press quenched direct and indirect extrusion.
(a) Billet temp.= 400°C, lnZ = 29.84, centre position
(b) Billet temp.= 300°C, lnZ_c= 27.10, peripheral location, direct extrusion
(c) Billet temp.= 400°C, lnZ_c= 25.87, centre location, direct extrusion
(d) Billet temp.= 400°C, lnZ_c= 25.87, peripheral location, direct extrusion
(e) Billet temp.= 400°C, lnZ_c= 26.81, centre location, indirect extrusion
(f) Billet temp.= 400°C, lnZ_c= 26.81, peripheral location, indirect extrusion

the stored dislocation density achieves a magnitude such that recovery by cross-slip and climb occurs resulting in the formation of subgrains. This mechanism is consistent with the observation that subgrains are first observed at grain boundaries and at high strain regions near the die entry. As extrusion proceeds the deformation extends backwards to form the deformation zone and isolated subgrain outcrops are observed progressively further back in the billet. In the later stages of extrusion when steady state conditions obtain the generation and annihilation of dislocations are in balance so that recovery and subgrain formation ensure zero work-hardening effects. At this stage subgrains are equiaxed and constant in size and misorientation. It should be recalled that in the earlier stages of deformation subgrain walls are ragged and elongated in the extrusion direction and the cells contain a high internal dislocation density. At peak pressure the steady state deformation zone is not fully established but this condition is achieved by the time the pressure has diminished by an amount equal to the additional increment required. Considering the necessary fact that changes in substructure require finite time intervals it is clear that the pressure peak is required to effect the establishment of the quasi-static deformation zone. A side effect is quite obviously that a quasi-static dead-metal zone is also formed. Hence it is not surprising that the pressure peak is associated with an activation energy similar to that for bulk self-diffusion and for creep since the climb of dislocations is the controlling process in each case.

Variation of substructure in the Extruded Product (Experimental Alloy 2014)

The variations in subgrain size in the transverse and longitudinal directions in the extrudate have been examined for direct and indirect extrudes processed at 300°C and 400°C with identical mean strain rates; the results are shown in Table 2 and illustrations of the substructures at centre and periphery in Fig.17. Note that the differing temperature compensated strain rates shown in Fig.17 are because of the differing temperature rises experienced in each mode.

At a position 0.25e (0.25 of the total length) from the front of the extrude, the results in Table 2 and micrographs c and d show that for direct extrusion there is a gradual increase in subgrain size in the transverse direction from centre to periphery which is consistent with the results presented above. The non-uniform flow in the direct process produces greater deformation at the periphery, inducing greater temperature rises and hence larger subgrains. This temperature rise may induce recrystallisation nucleation in very short time periods and this is illustrated in Fig.17b which is of a specimen extracted from the mid-radius position at 0.25e in the 300 C direct extrudate. Thus although the extrudate was press quenched and the billet temperature relatively low a recrystallisation nucleus can clearly be distinguished. Micrographs e and f in conjunction with Table 2 verify that the transverse variation in the indirect extrudates is minimal confirming a smaller temperature variation across the product. Comparison of the second phase particles in micrographs e and f also confirms that there is a difference in morphology which develops transversely in the extrudate due to minor inhomogeneities of flow in the indirect process. The variation in subgrain sizes at 0.3e, 0.6e and 0.8e in the extrudes processed with an initial billet temperature of 300 C is also shown in Table 2. In the case of direct extrusion the subgrains at the centre of the extrude progressively increase in size as extrusion proceeds until at 0.81 there is only a minor variation in the transverse direction. This observation is consistent with most temperature models which would predict an increase in temperature with a transverse gradient gradually decreasing because of the high thermal conductivity of aluminium alloys. The indirect extrusion shows only minor increases in subgrain size as extrusion proceeds and in general a much more homogeneous extrudate is produced.

Modification of Substructure during Solution Treatment

The standard solution treatment for F-temper 2014 alloy is 20 minutes to 2 hours at 500°C (dependent upon size) followed by water quenching to obtain a supersaturated solution. Fig.18 shows optical micrographs of specimens soaked for varying times at 500°C and subsequently cold water quenched. In Fig.18a and b are shown the structure in the transverse and longitudinal directions, respectively, after a solution soak of 0.5 hours. The micrographs suggest that the extruded, fibrous structure has been replaced by large grains which are elongated in the extrusion direction. Clearly, recrystallisation has occurred, but the amount of nucleation seems to have been limited.

In order to examine the effect of the extrusion-process parameters on the solution-soak sequence, specimens were prepared from material extruded over a wide range of temperatures, which was exposed to a 1 hour, 500°C soak and examined in the electron microscope. Fig.19a shows the substructure of an extrudate processed at a temperature compensated strain rate Z of 5 x $10^{10}s^{-1}$ (initial billet temperature 300°C) which was typical of all high Z extrusions. The structure is fully recrystallised with the particles aligned in the extrusion direction; there was no evidence of residual subgrains.

At lower values of Z (9 x 10^9 s^{-1} initial billet temperature 475°C) the typical substructure morphology was as shown in Fig.19b which, although in the solution-treated condition, verified that subgrains may be retained throughout the solution treatment. This observation was unexpected because the optical microscopy reported above had indicated that this alloy recrsytallised under all processing conditions. The proportion of retained substructure was found to increase with decreasing Z, suggesting that at higher extrusion temperatures there is very little driving force for recrystallisation. The grain size in the solution treated material decreased with decrease in extrusion temperature at constant ram speed and extrusion ratio as indicated in Table 3. The table also shows that grain sizes are smaller after direct than after indirect extrusion.

The measurements at high temperature should be viewed with some circumspection because of the retained substructure observed in the electron microscope. Clearly, recrystallisation at higher temperatures is incomplete. This feature was investigated further

146

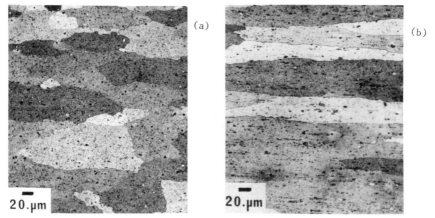

FIGURE 18. Microstructure of billets after 500°C solution treatment
with water quench. (a) 0.5 hours soak, transverse section
(b) 0.5 hours soak, longitudinal section.

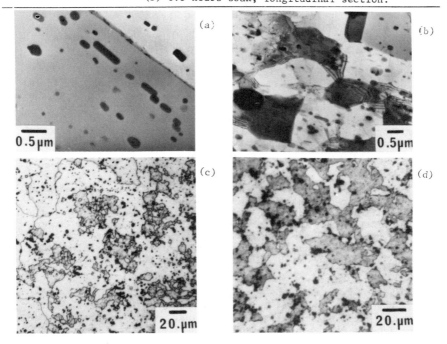

FIGURE 19. Extrudate microstructure following post-extrusion soak,
500°C, 1 hour, and quench.
(a) Extrusion temperature 300°C, $Z = 5 \times 10^{10} s^{-1}$, TEM
(b) Extrusion temperature 475°C, $Z = 9 \times 10^9 s^{-1}$, TEM of
typical substructure
(c) As b, optical micrograph, retained substructure
(d) As b, optical micrograph, partially recrystallised
duplex substructure

147

by etching the optical specimens in Ramsey's reagent as an alternative to the modified Keller's reagent used previously. This revealed the presence of a duplex structure (i.e. partially recrystallised) and a retained substructure as shown in Fig.19b and 19d.

Effect of Age-Hardening on Structure

The previous section has illustrated the structural changes which occur during the solution-soak treatment and that the structure after treatment is to some extent dependent upon the extrusion-process variables. Hardness measurements showed that the material softened considerably during solution treatment and, despite the structural differences observed, was constant for all of the extrudates. The retained and partially-retained substructures do not seem to contribute to strength. The strength of the alloy can, however, be considerably improved by natural ageing at elevated temperatures. This strength increase may be attributed to the formation of GP1 and GP B zones which act as obstacles to dislocation motion. Clearly, the greater the amount of solute in solution, the greater the influence of these zones.

Fig.20 shows electron micrographs of material solution treated and then aged for 18 hours at 160°C. The most important structural feature directly related to the extrusion process is the retained substructure, shown in Fig.20a for an extrusion performed at $Z = 5 \times 10^9 s^{-1}$. The subgrains remain well formed with clean boundaries and low misorientations: a fine dispersion of precipitates is also evident. This fine precipitation is shown at higher magnification in Fig.20b, revealing the formation of θ'' platelets throughout the structure. At this magnification it is impossible to identify a precipitate-free zone associated with the subgrain boundary, and it was only at very high magnification that this recognized feature of aged aluminium alloys could be detected. This micrograph also indicates that preferential precipitation has occurred at the subgrain boundary. It seems unlikely that these particular features will have any substantial influence on the properties of 2014 alloy.

The elastic-coherency strains associated with the θ'' zones give rise to contrast on the electron microscope and Fig.20c shows that the strain fields extend from one precipitate to the next, forming a checkered pattern. As the strains near θ'' plates become larger, they may be relieved by the formation of an extra plane of atoms parallel to the θ'' plane, thus producing a stable dislocation ring. This can also occur at θ'' plates, and such rings can readily be identified in Fig.20c.

However, the strains due to dislocations near the θ'' plates are localised at the interface and do not spread into the matrix, as is the case for elastic-coherency strains. It is easier for dislocations to move through a structure containing partially-coherent precipitates, hence overageing will result in strength and hardness reduction. Conversely, overageing is likely to produce better fracture-toughness and corrosion properties.

148

TABLE 2. Subgrain Size Measurements in the
Transverse and Longitudinal Direction of
Direct and Indirect Extrudes of 2014 Alloy.

Direct: $T_i = 300°C; R = 30:1; v = 8$ mm/s.

Distance Along Extrude		Centre (μm)	±	Mid-Radius (μm)	±	Edge (μm)	±
0.25	long	1.59	0.22	1.68	0.12	1.75	0.15
	trans	1.05	0.19	1.07	0.13	1.22	0.18
0.6	long	1.67	0.28	—	—	1.71	0.19
	trans	1.17	0.11	—	—	1.28	0.11
0.8	long	1.70	0.36	—	—	1.77	0.33
	trans	1.25	0.11	—	—	1.43	0.16

Indirect: $T_i = 300°C; R = 30:1; v = 6.5$ mm/s.

Distance Along Extrude		Centre (μm)	±	Mid-Radius (μm)	±	Edge (μm)	±
0.25	long	1.58	0.07	1.62	0.41	1.66	0.12
	trans	0.96	0.18	0.87	0.08	1.14	0.20
0.6	long	1.62	0.15	—	—	1.71	0.31
	trans	0.89	0.15	—	—	1.12	0.30
0.8	long	1.62	0.06	—	—	1.68	0.28
	trans	1.09	0.06	—	—	1.13	0.11

Direct: $T_i = 400°C; R = 30:1; v = 8$ mm/s.

Distance Along Extrude		Centre (μm)	±	Mid-Radius (μm)	±	Edge (μm)	±
0.25	long	2.51	0.21	2.70	0.50	2.87	0.76
	trans	1.70	0.31	1.74	0.15	1.84	0.30

Indirect: $T_i = 400°C; R = 30:1; v = 6.5$ mm/s.

Distance Along Extrude		Centre (μm)	±	Mid-Radius (μm)	±	Edge (μm)	±
0.25	long	2.44	0.31	2.45	0.13	2.26	0.37
	trans	1.82	0.25	1.49	0.10	1.76	0.27

TABLE 3. Extrudate Grain Size after
Solution Treatment.

Extrusion Mode	Billet Temperature °C	Solution-Treated Grain Size, μm
Direct	480	36.0
Direct	440	33.4
Direct	400	31.3
Direct	350	26.8
Direct	300	22.3
Indirect	480	62.0
Indirect	440	62.0
Indirect	400	62.0
Indirect	350	49.4
Indirect	300	39.5

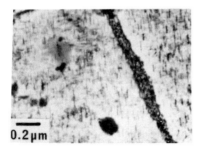 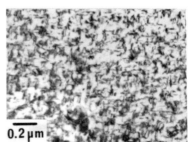

FIGURE 20. (a) Retained substructure with fine precipitates
(b) Subgrain boundary in retained substructure
(c) θ" precipitates.

Relationship Between Structure and Properties

The discussion above has indicated that the structure of the extrudate is dependent on the total thermomechanical cycle but that the most important consideration in terms of properties is the subgrain size which very much determines the metal response to the solution and ageing sequence. The reader should recall that the subgrain size is related to the processing parameters and has been established for AA 2014 by Paterson (9) to be :

$$d^{-1} = 0.096 \ln Z - 1.747 \text{ for direct extrusion}$$

$$d^{-1} = 0.085 \ln Z - 1.586 \text{ for indirect extrusion}$$

In these formulae d is the subgrain diameter and Z is the temperature compensated strain rate. Thus once the homogenisation and preheat sequence have been completed structure control is largely dependent upon the extrusion conditions. The discussion above has also indicated that there is a variation of structure longitudinally in the extrude which could and should be eliminated by the control of press speed during the cycle.

The T6 temper is the most commonly used heat treatment for 2XXX alloys and has therefore been investigated for direct and indirect extrusion. The extrusions were subjected to a 30 minute soak at 550 C, quenched and subsequently aged at 160° C for 18 hours. The tensile properties for these billets subjected to homogenisation and a one step reheat are shown in Fig.21 for both direct and indirect extrusion. Fig.21 also shows T6 (press quenched) properties. It is clear that there is a discontinuity in the proof stress and ultimate stress values which is not the result of any gradual variation of extrusion temperature. The high strength or low Z regions correspond to the unrecrsytallised substructures shown in the previous section whilst the low strength levels are associated with completely recrsytallised structures. The problem then is to ascertain whether the increase in strength is due solely to substructural strengthening or is a result of the preferential precipitation occurring at subgrain boundaries. Recent work (17) has shown that for AA 2014 alloy both high and low temperature extrudates exhibiting retained substructures and fully recrystallised structures respectively have identical ageing characteristics indicating that although preferential precipitation does occur it has little effect on the attainment of peak strength.
This is perhaps to be expected since the evidence shown above indicates that the amount of heterogeneous precipitation at subgrain boundaries and dislocations is not as great as the precipitation within subgrains. Hence we would expect the strength to vary with ln Z and this indeed can be detected in the figure although there are insufficient data points to draw firm conclusions. At higher values of Z the strength increases presumably due to finer grain sizes but never achieves the levels possible when processing using low Z conditions. It is interesting to note that there are differences in the variation of properties between direct and indirect extrusions: the indirect extrusions showing less variation in properties. This can be directly attributed to the more homogeneous nature of the structure produced when proceeding by the indirect route. Thus we may conclude that

150

as far as tensile properties are concerned there is some advantage in processing at as low a Z as possible consistent with the attainment of a satisfactory surface. The critical nature of surface recrystallisation may be alleviated by producing by the indirect method.

One of the most limiting factors of the extended application of high strength aluminium alloys is their relatively low fracture toughness. The structural information presented above indicates that since the structural morphology is radically altered by the prevailing conditions, then so should the fracture path. Fig.22 shows that the K_{1C} value as a function of Z decreases by twenty percent from which it can be seen that the processing effects on fracture are somewhat dramatic. The results are for 2014 alloy in the T6 condition (scaked 500°C for half-an-hour, aged at 160°C for 18 hours). Evaluation of fracture properties was by the crack opening displacement method specified in BS 5762:1979. The values of K_{1C} are of the same magnitude reported by other workers and it is apparent that the best fracture toughness properties are produced in the material processed under low Z conditions. The rapid decrease in fracture toughness with increasing Z is greater than can be attributed to the corresponding increase in the yield stress and is contrary to the expected increase in fracture toughness due to the observed decrease in recrystallisd grain size with increasing Z. It is thus evident that the fracture toughness must be associated with other factors. The fracture surfaces as examined under the scanning electron microscope are shown in Fig.23. Optical examination had previously indicated that the low Z specimens appeared to exhibit smooth ductile characteristics whilst high Z extrusions indicated a faceted fracture. Since all the specimens received the identical solution soak and age sequence the fracture mode is clearly associated with the extrudate rather than the heat treatment. Fig.20 has indicated that a substructure may be retained through a heat treatment schedule and Fig.23 indicates that although the facets are intergranular in appearance for the high Z(a and b) extrude the facet surface is not smooth. These surfaces are shown more clearly in Fig.23 which shows the fracture to be ductile but the dimples are large and deep suggesting that the fracture was predominantly intergranular. The presence of relatively large grain boundary precipitates observed in the structure of the low temperature extrude also suggests that the mechanism of fracture proceeded by the nucleation of voids at the grain boundary particle/matrix interface. Clearly the dimple size will be related to the size of the precipitates which explains the variation observed in the micrograph. The fracture surface observed in high temperature extrudates is shown in Fig.23c and 23d and it is evident that the fracture is transgranular. The observed dimples are 1-2 μm in size and the similarity between the size and spacing of the dimples and the coarse second phase particles suggest that fracture is nucleated at those particles by decohesion of the particle/matrix interface. The fracture surface of the specimens extruded at intermediate Z's tended to show mixed fracture characteristics in proportion to the retained substructure. The fracture mode thus progressively changes from intergranular to transgranular as the extrusion process parameters are varied from high to low Z. Clearly the energy required for the formation of a void at a grain boundary thus leading to lower toughness in the high Z produced extrudate. We

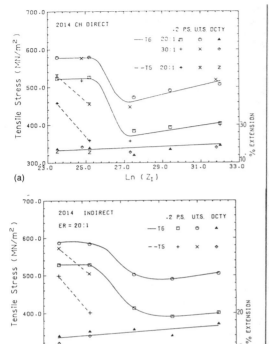

(a)

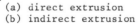

(b)

FIGURE 21. Tensile properties of T6
    temper products.
        (a) direct extrusion
        (b) indirect extrusion

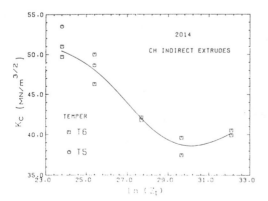

FIGURE 22. Fracture toughness variation
with temperature compensated strain
rate for indirect extrusion.

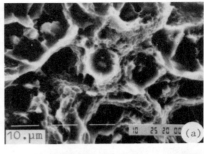

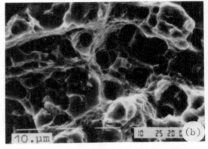

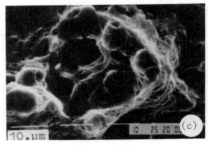

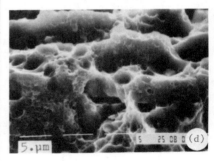

FIGURE 23. Fracture surface of
    extrudates subcutaneous to the
    fracture surface.
    (a),(b) high Z extrudes
    (c),(d) low Z extrudes.
    Direct and Indirect
    Extrudes respectively.

may conclude that the presence of retained subgrains significantly increases the fracture properties compared to those of a finely recrystallised material.

Conclusions

A prolonged homogenisation treatment is required for the production of a suitable extrudate structure in aluminium alloys; the extrudability may be influenced and improved by the preheating operation. The development of substructure spreads from the die entry region and is influenced by second phase particles and by grain boundaries. The substructural development is more homogeneous in indirect extrusion but the rate of change of structure is much greater in the direct case such that consistency of product may be easier in the case of direct extrusion. It is shown that an additional increment of pressure is required to establish the deformation zone and this may be directly related to the substructural activity. The substructure is determined by the processing parameters and can be important even when the product is subjected to a solution soak and ageing process. Indeed under suitable processing conditions the substructure can be partially retained after a solution soak sequence.

Acknowledgements.

The authors would wish to express their appreciation of continued support from ALCAN INTERNATIONAL who supplied the material used in this work and considerable financial backing. Student support has also been supplied by the Science and Engineering Research Council and this is also acknowledged.

References

1. Bird, J.E., Mukherjee, A.K. and Dorn, J.E., 'In Quantitative Relation Between Properties and Microstructure' (D.G. Brandon and A. Rosen, Eds.), 1969, pp.255-342, Isr. Program Sci.Transl. Jerusalem.

2. McQueen, H.J., Wang, W.A., and Jonas, J.J., (1967), Can.J. Phys., 45, 1225 - 1235.

3. Sellars, C.M. and Tegart, W.J.McG.)1972), Int.Met.Rev., 17, 1-24.

4. Gagna, R.G. and Jonas, J.J., Trans.Met.Soc., AIME, 1969, 245, 2581.

5. Raybould, D. and Sheppard, T., J.I.M., 1973, 10', 45.

6. Castle, A.F. and Sheppard, T., Met.Techn. 1976, 3, 454.

7. Sheppard, T., Proc. 2nd Int.Sess. Al.Ext. Techn. 1977, 1, 337, publ. Al. Association.

8. Tutcher, M.G., Ph.D. Thesis, University of London (Imperial College), 1979.

9. Paterson, S.J., Ph.D. Thesis, University of London (Imperial College), 1981.

10. Vierod, R.P., Ph.D. Thesis, University of London (Imperial College), 1983.

11. Wood, E.P., Ph.D. Thesis, University of London (Imperial College), 1878.

12. Castle, A.F., and Sheppard T., Met.Tech. 1976, 3, 492.

13. Humphreys, J.H., Acta Metall. 1977, 25, 1323.

14. Sheppard, T., McShane, H.B., and Tutcher, M.G., Powder Met, 1978, 21, 47.

15. Smallman, R.E., 'Modern Physical Metallurgy', 1970, London, Butterworths.

16. Nicholson, R., Thomas, G, and Nutting, J.; J.I.M., 1958, 87, 429.

17. Sheppard, T., Proc. 3rd Int.Sess. Al.Ext.Techn. 1984, 1, 107, publ. Al. Association.

DYNAMIC RECRYSTALLISATION IN ALUMINIUM ALLOYS

T.Sheppard, M.A. Zaidi, M.G. Tutcher, N.C. Parson

Department of Metallurgy and Materials Science,
Imperial College of Science and Technology
London, S.W.7.

Summary

The deformation by both extrusion and rolling have been investigated for alloys containing 5%, 7% and 10% magnesium. The 5% and 7% alloys contained Mn giving a precipitate dispersion whilst the 10% alloy was a powder alloy containing oxide dispersions. It is shown that dynamic recrystallisation is a feature of the deformation mode of each alloy; becoming increasingly prevalent as the Mg content increases. The mechanism is however significantly different from 'classic' dynamic recrystallisation; it does not involve grain boundary bulging. Nuclei are formed at sites of high dislocation density; either second phase particles or grain boundaries. There is no evidence that the increased Magnesium content acts to lower the stacking fault energy of Aluminium Alloys.

Introduction

It is generally accepted that when aluminium is deformed the original grains are preserved with a modified morphology and the deformation occurs solely by dynamic recovery thus introducing a substructure into the individual grains (1), (2). It is also accepted that most metals and alloys exhibiting a high stacking fault energy deform by a similar mechanism whilst those of low stacking fault energy deform by dynamic recrystallisation associated with a grain boundary bulging mechanism. (3) - (6). The decreased stacking fault width in such alloys impedes dislocation mobility thus modifying the balance between dislocation generation and annihilation. The resulting increase in dislocation density at the grain boundaries nucleates the recrystallisation process.

Truszkowski (7) and others (8), (9) have suggested that the addition of 1% Mg to Al reduces the stacking fault energy from 200 to about 50 J m^{-2} and hence dilute Al-Mg alloys would be expected to dynamically recrystallise. However considerable published information on Al alloys containing 2% Mg (10) and 3% Mg (11) indicates that in these alloys dynamic recovery is the only softening mechanism.

More recently McQueen et al (12), Sheppard and Tutcher (13), Sheppard, Parson and Zaidi (14), have each shown that alloys containing greater than about 4.5% Mg exhibit a duplex deformation mode which includes dynamic recrystallisation at least when deformed to high strains. The effect appears to become more pronounced as the Mg content increases but is associated with the accumulation of strain at large particles rather than with grain boundary bulging.

This communication compares the dynamic recrystallisation occurring in alloy AA 5456 with higher Mg bearing alloy containing 7% Mg and 10% Mg and deformed at both low and high strains.

Experimental Details

The materials, supplied by Alcan International Ltd. (AA 5456 and 7% Mg) and by Alcoa (10% Mg), had the chemical compositions shown in Table 1. The AA 5456 alloy and the 7% g alloy were supplied in the form of DC cast logs whilst the 10% Mg was supplied in powder form having a mean particle size of 22μm. The powder was cold compacted into billets 75mm diameter x 120mm long and a density of 86% prior to extrusion.

Table 1.   Chemical Composition of Alloys

Designation	Mg	Mn	Fe	Cr	Si	Zn	Cu	B (ppm)	Al
Powder 10Mg	10.4	0.01	0.07	-	-	0.39	-		Rem
Al-7g	7.4	0.68	0.17	0.17	-0.06	0.005	0.003	17	Rem
A65456	5.1	0.7	0.18	0.12	0.06	0.005	0.001	12	Rem

The cast billets were homogenised at 500 C for 24 hr. They were then reheated to 450 C by induction transferred automatically to the press container (maintained at 375 C) and extruded at a rate of 40:1. Experimental details of the extrusion process and specimen preparation may be found in references (10), (11), (13) and (14).

Results and Discussion

The macrostructural features of the AA 5456 billets partially extruded to locations 1-5 indicated in Fig.1 were similar to those described in a companion paper also presented at this conference and will not be further discussed. The location of specimens for electron microscopy is shown in Fig.2.

Microstructural Observations (AA 5456 alloy)

Very little substructural activity could be detected in the specimens taken from the billet arrested at location 1 in the extrusion cycle. Isolated outcrops of dislocation walls were observed in the die-entry region (A) but otherwise the cast grains appeared to contain only residual dislocations. The dislocation walls observed were associated with high-angle boundaries or second-phase particles. Compared with the superpure alloy investigation (15) the deformation appeared to be homogeneous, which indicates the importance of grain size on deformation. When the extrusion was interrupted at location 2 a more highly deformed structure was observed around the die-entry region. The deformation becomes sufficient to distort the original grain structure and outcrops of subgrains could be observed at locations A and B within the billet. Elsewhere, dislocation density was greatest at second-phase particles and high-angle grain boundaries, the effect of second-phase particles being to modify the dislocation distribution by pinning such that walls are produced to form triangular subgrains typical of those shown in Fig.3a. In the regions in which deformation was more advanced subgrains appeared as a banded substructure (Fig.3b) in which the contrast can be seen to suggest higher misorientations laterally than longitudinally. In general the longitudinal direction corresponded to the direction of deformation suggested by the second-phase particles. The enhanced subgrain formation appears to be associated with the larger second-phase particles which would in any case permit greater lattice rotations to be generated. This specimen did, however, contain regions in which subgrain activity could not be detected, indicating incomplete establishment of the deformation process. Specimens taken from location 3 indicated the formation of a dislocation substructure at points A, B and C, although in the rear of the billet there were still regions in which dislocation activity was negligible.

In most regions, however, subgrain walls, typically as shown in Fig.3a, were observed with second-phase particles appearing to play an important role in determining deformation: pinning both subgrain and higher angle boundaries. Fig.4a is a micrograph of one of the regions exhibiting little grain distortion. The only evidence of the applied stresses is the appearance of weak sub-boundaries and dislocations pinned by second-phase particles and contained by deformation or transition bands within the grains. These are the results of strain inhomogeneity during compression and the early stages of distortion. Localized lattice rotations are produced

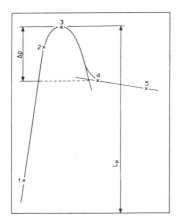

FIGURE 1. Location of sampling positions on load-displacement locus.

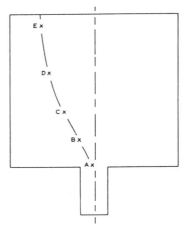

FIGURE 2. Location of electron
microscopy samples.

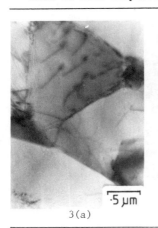

3(a)

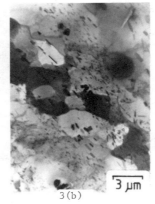

3(b)

FIGURE 3.
(a) subgrain walls
pinned by second-phase
particles.
(b) subgrain band
at location 2A.

(a)

(b)

(c)

FIGURE 4. (a) typical subgrain formation; (b) typical subgrain formation
(c) deformation zone corresponding to peak pressure.

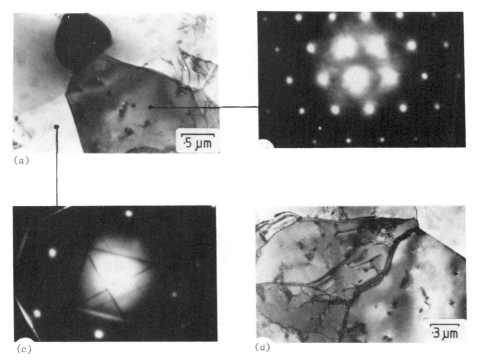

(a)

(c)   (d)

FIGURE 5. (a) dynamically recrystallised grain at a second-phase particle
          (b) and (c) SADP's from a
          (d) dislocation activity within recrystallised grain in (a).

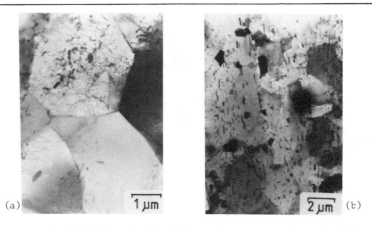

(a)   (b)

FIGURE 6. (a) Subgrains in deformation zone at location 4A.
          (b) Typical subgrain formation at location 5C
              showing particle alignment.

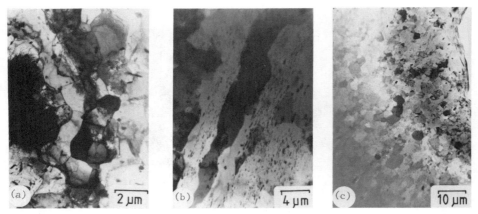

FIGURE 7. (a) Equiaxed subgrains at location 5B; (b) Subgrain banding in the die mouth region; (c) Substructure within deformation zone.

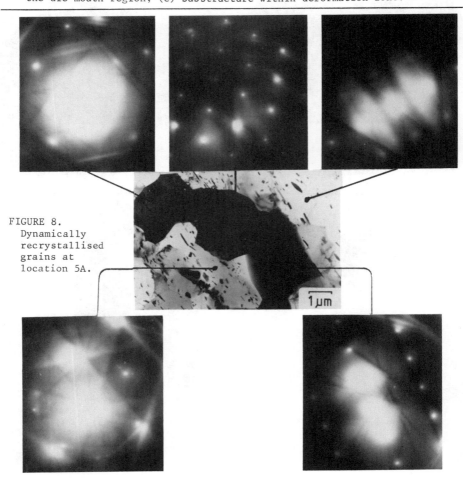

FIGURE 8. Dynamically recrystallised grains at location 5A.

resulting in the optical contrast shown in the figure. This in turn implies an enhanced dislocation generation to accommodate strain gradients and explains the inhomogeneous subgrain structures observed in those regions. The dislocation substructures at locations towards the die exit (A, B, C) at location 3 were typically as shown in Fig.4b. The subgrain size is heterogeneous and apears to be due to the inhibition to movement caused by the precipitate distribution. At locations A and B the microstructure is significantly different from that found in the superpure material previously examined. The optical micrograph of this area shown in Fig.4c indicates the original grains to be aligned in the deformation direction with small equiaxed grains observable at the original grain boundaries. Such grains were not observed in the superpure material and must have been produced either statically during the short time between deformation and quenching or they are evidence of a dynamic recrystallisation process occurring during deformation.

Fig.5a shows what appears to be a dark subgrain associated with a precipitate particle of about 0.5μm. The selected-area diffraction patterns (SADP's) of it and its nearest neighbour (Fig.5b and c) indicate that the misorientation across this boundary is about 45°, hence this is certainly a high-angle rather than a low-angle boundary. The size of this grain or subgrain is of the order of 1.5μm which exempts its consideration as an original grain: it must therefore be the result of a recrystallisation process. Further investigation (Fig.5d) reveals an extensive dislocation network within the grain consisting of tangled high-density areas and extensive pinned dislocations: evidence that the grain was formed dynamically since the sweeping of a high angle boundary during a static recrystallisation process would have removed such dislocations. Hence, the nature of the equiaxed grains identified in Fig.4d would appear to be due to dynamic recrystallisation. Specimens examined from location 4, corresponding to a pressure decrease of p from the 'breakthrough pressure' showed well developed substructures at locations A, B and C as well as increased dislocation activity towards the rear of the billet. The continuous subgrain structure identified in the die mouth was heterogeneous, the subgrain size being smaller at high-angle boundaries where evidence of a high internal dislocation density may also be observed (Fig.6a). This is consistent with poor recovery combined with greater deformation at these locations. Optical microscopy revealed the presence of small equiaxed grains at original boundaries similar to those observed at the peak pressure location (Fig.4c). Samples extracted from the steady-state region (location 5) showed the existence of subgrains at all locations in the billet. At the rear of the billet subgrains were located mainly at high-angle boundaries, further proof of the role of grain boundaries in causing preferential heterogeneous deformation at points of maximum strain. In the deformation zone region the substructure showed more homogeneity with transboundary contrast indicating low misorientation between subgrains. A typical subgrain formation at location C is shown in Fig.6b, which illustrates that second-phase particles are aligned into the deformation direction at an early stage in the deformation process. Further forward at location B the substructure may be observed to be equiaxed (Fig.7a) and to contain a high internal dislocation density indicating the inability of the material to completely recover. At location A subgrain banding was observed (Fig.7b) a feature found in many locations in the superpure material. The cause of banding could be due to local differences in stress conditions because with this fine-grained material and at this location the presence of small recrystallised grains could create a complex stress condition. It is also possible that any one pair of grains may be constrained to deform in the extrusion direction only thus supporting the bulk deformation theory. Since there is no evidence of dynamic recrystallisation being associated with the banding phenomena the latter explanation would appear to be more likely. Fig.7c shows a low magnification electron micrograph from location A in which the distribution of subgrains can clearly be seen to be generally homogeneous and of low contrast. There are however numerous 'subgrains' which are dark in appearance suggesting higher misorientations. A higher magnification micrograph of one such area is shown in Fig.8 together with the appropriate SADP's. There are three 'dark' subgrains and their misorientations to each other and to adjacent 'light' subgrains vary from 25° to 45° proving that they must be recrystallised grains. The presence within the grains of a tangled dislocation substructure proves that they have been formed by a dynamic process. Many such areas were identified in the deformation region showing that in this alloy the deformation mechanism is modified and that it is duplex consisting of both dynamic recovery and dynamic recrystallisation. The effect of this observation on properties is significant and will be discussed below.

Extrudate Structure

The duplex softening mechanism operating during deformation has a very profound effect upon product structure. In Al alloys in which dynamic recovery is usually the only mechanism operating the extrudate normally would have a substructure throughout the material which might be modified by a recrystallised layer at the periphery if the strain, temperature and strain-rate conditions were suitable to create the necessary driving force. In commercial extrusion the increased strain at the surface and typical process conditions usually combine to ensure that recrystallisation nucleation does occur, but only at the surface of the extrudate. In the case of the extrusion of the commercial Al-Mg alloy investigated the surface appeared recrystallised over a wide range of processing parameters (Fig.9). The depth of the recrystallised layer varied linearly with the logarithm of the temperature-compensated strain rate Z being greater as Z decreased (16) i.e. as the temperature increased. The recrystallised grain size was also much smaller as Z increased. The interior of the material consisted of a dynamically recovered structure modified by the presence of equiaxed dynamically recrystallised grains at the original grain boundaries. These grains have a profound effect upon product tensile properties which normally can be directly related to Z and exhibit decreasing strength and increasing ductility with progressive raising of the Z parameter.

Development of dislocation substructure

Dislocation generation commences and spreads from the die-entry region because it is in those regions that strain and strain gradients are greatest. The activity appears to be greatly assisted by high dislocation densities associated with non-deformable second-phase particles which introduce additional sites for subgrain formation and subsequently effectively pin the structure. Before the peak pressure was obtained dislocation density increased rapidly as a consequence of the interaction of mobile dislocations with each other and with

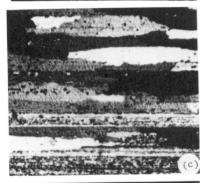

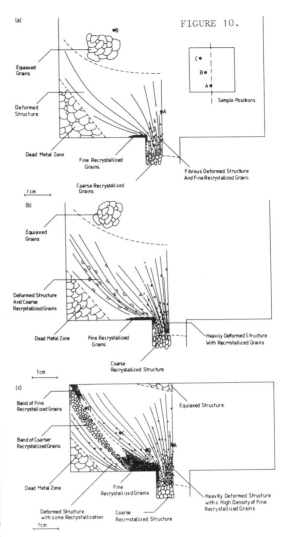

FIGURE 10. Schematic of structural
events in deformation zone,
Al-7Mg alloy.

FIGURE 9. (a) peripheral structure
R = 10:1, T = 595K, V = 5.4 mm s^{-1}, x 64;
(b) internal structure, same values
of R, T and V as in (a), x 128;
(c) peripheral structure, R = 10:1,
T = 735K, V = 12.6 mm s^{-1}, x 64;
(d) internal structure, same values
of R, T and V as in (c), x 64.

dislocations stored within original grains. Dislocation density increases most rapidly in the regions in which strains and strain gradients are highest, thus grain boundaries located in the deformation zone are prime activity sites. Before the attainment of peak pressure the stored dislocation density has achieved a sufficient magnitude such that recovery by cross-slip and climb occurs, resulting in the formation of subgrains. Recovery is incomplete, however, since the subgrains have a high internal dislocation density. Following the attainment of peak pressure a critical strain for dynamic recrystallisation is achieved in the forward regions of the billet. The fine subgrain structure developed before the onset of dynamic recrystallisation implies that nucleation by grain boundary bulging is unlikely because the length of unpinned boundaries are relatively short. It should also be recalled that the duplex substructure is observed only in regions where significant deformation has occurred: there is no evidence that pre-existing grain boundaries are the principle nucleation sites for dynamic recrystallisation by a 'bulge' mechanism. It would therefore appear, from the evidence presented above, that nucleation occurs by preferential increased misorientastion in selected favourable subgrains. Nucleation would thus be more likely to occur in regions adjacent to grain boundaries where the strain gradients are greatest in order to accommodate plastic anisotropy. A further preferential site is at second-phase particles where the dislocation density is greatest. Elsewhere, the critical strain does not seem to have been substantially attained because only outcrops of dynamically recrystallised grains can be observed. As extrusion proceeds the deformation extends back into the billet to form the quasistatic deformation zone so that there may even be a decrease in mean strain and strain rate when the pressure has dropped by the amount p. At this point subgrain outcrops are observed further back in the billet and the subgrain size appears to increase slightly. Deformation in the steady-state region still occurs by a duplex mechanism with dynamic recrystallisation generally being observed only close to original grain boundaries or at second-phase particles: it is only at sporadic points that such a mechanism appears to have occurred because of a random misorientation which could be attributed to high dislocation densities within subgrains. There is no evidence that dynamic recrystallisation occurs because of a modification to stacking-fault energy. The evidence suggests a much simpler explanation: it is the solute addition and associated vacancy redistribution which is the prime cause of the modification to a duplex deformation mechanism, and precipitates and second-phase particles then assume the necessary secondary role.

Macroscopic Examination (Al-7Mg Alloy)

The structural events occurring during the establishment of quasistatic flow were observed by polarized light microscopy and are presented in schematic form in Figs. 10a-c. At breakthrough pressure (location 3 in Fig.1) illustrated in Fig.10a, deformation was found to extend well back into the billet while the dead metal zone (DMZ) and the usual shear zone adjacent to it were not very well defined. The most interesting feature was the presence of very fine grains (5-10 m dia.) in the intense deformation zone adjacent to the die orifice and located in the die/metal interface. The initial grain size of this alloy was established as 115µm, so these fine grains were obviously the result of recrystallisation during and/or after extrusion. Some

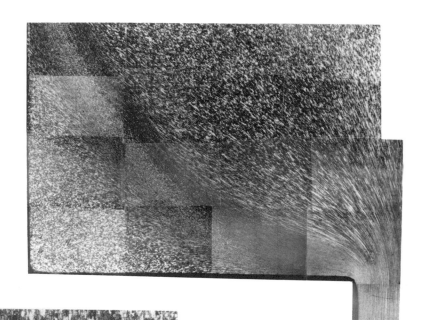

FIGURE 11(a).

Composite micrograph of
partially extruded billet
at location 5 of Fig. 1.

FIGURE 11 (b).
(i)   Core of partial extrude 1 cm
      from die exit;
(ii) Surface region of extrude
      1 cm from die exit;
(iii) Microstructure taken at
      location 3 of Fig.1 of
      a core of partial extrude
      at die exit.

fine grains were also observed within the deforming region as far as 15mm back into the billet from the die face and generally associated with grain boundaries, indicating that grain boundaries, or particles at grain boundaries, were acting as preferential nucleation sites. At this position, extrusion had commenced and the structure consisted of mixed fibrous and fine equiaxed grains in the die bearing area, followed by coarser and often elongated grains in the remainder of the extrudate. Thus, at this stage, the structure observed microscopically was quite different from that observed under identical conditions in an Al-5Mg alloy.

The extrusion cycle was next terminated at location 4 (Fig.10b). The diagram indicates that the deformation zone had considerably expanded: it was observed that the DMZ was now well defined and also had decreased in size. The shear zone adjacent to the DMZ was more intense and both fine and relatively coarse grains were observed alongside the original heavily deformed grains. The density of fine crystallised grains both adjacent to the die and within the deformation zone had increasd considerably. Figure 10c illustrates the macroscopic features of metal flow at location 5, Fig.1, while Fig.11a shows a composite micrograph taken from the same location of most of the partially extruded billet. The small recrystallised grains identified earlier in the extrusion process can be seen to be present throughout the deformation zone. In the areas of most severe deformation (i.e. adjacent to the DMZ and in the die mouth region) both fine and coarse recrystallised grains are present but the morphology of these regions is dominated by the high density of very small recrystallised grains.

Figure 11 c illustrates the structural differences which may be observed when moving from the die face back into the billet along a line corresponding to the inner limits of the DMZ. On the left, adjacent to the die/metal interface (lower edge in Fig. 11 c), the original grains can be seen to be equiaxed, and on the right (top) they are beginning to elongate as they flow towards the die. A few very fine recrystallised grains and particles that appear to be recrystallisation nuclei can be seen located, in general, at the original grain boundaries. Nearer to the shear zone (i.e. towards the top of the micrograph) the original grains have become elongated and the density of small equiaxed recrystallised grains significantly increases. The shear zone is clearly of a duplex nature containing greatly elongated grains of a fibrous nature commonly observed in Al alloys in combination with a large number of both fine and and coarser recrystallised grains.

Finally, at the right hand side upper extremity of the micrograph the structure appears to be almost fully recrystallised but with a banding effect, such that the original grain boundaries are readily discernable. Fine recrystallised grains were observed at the original grain boundaries in the Al-5Mg alloy, but clearly the density of such grains in the Al-7Mg alloy, processed under identical conditions, is at least two orders of magnitude larger. This would appear to indicate that the increased Mg content leads to considerable enhancement of microstrain accumulation, thus producing large numbers of potential nucleation sites. Another important feature which can be observed clearly in Fig.11a is that the recrystallised grains are present both as very fine ($> 5\mu m$) and relatively coarse ($< 25\mu m$) grains. This suggests that some of the nuclei form during extrusion and grow preferentially during the delay between the cessation of deformation and the quench. Indeed, if dynamic recrystallisation does occur, it would be expected that, at any instant, some of the grains would be dislocation free, thus able to grow because of the energy difference between the grain and surrounding material while others, nucleated earlier in the deformation cycle, would already have attained a substantial substructure and, therefore, would tend to be stable. Fig. 11b illustrates the structure of the partial extrudate taken at location 5 of Fig.1. When the material has just emerged from the die land area [i], the core of the extrudate appears to have a typically fibrous structure, but inspection of the micrograph reveals that the fibrous appearance is deceptive and that the structure consists of deformed elongated grains and a very high density of very fine recrystallised grains.Fig 11[ii] shows that 1cm further away from the die the core structure had dramatically changed. Although the original fibrous nature can still be identified, the structure is now one of fine and coarse recrystallised grains and is almost completely recrystallised. Since the billet and extrudate were quenched simultaneously, the differences in structure observed between [i] and [ii] are remarkable in that they occurred in about 0.08s. This time interval is certainly much shorter than any reported work known to the authors could justify for purely static events to occur in Al alloys. The surface structure at the same position is presented in [iii] which shows the surface to consist of very fine recrystallised grains while subsurface layers contain progressively coarser recrsytallised grains. This observation would be consistent with the more highly strained surface containing a greater density of nuclei on exit from the die land area. The morphology is similar to that observed in other Al alloys (17) but the structural scale of the grains is an order of magnitude smaller.

Since it is possible to quench material much more rapidly after rolling than after partial extrusion, billets were prepared, hot rolled, and immediately quenched, thus avoiding the delay time in the experiments reported above. A thermocouple was inserted into the central axis of the billet, connected to a data logger, and the quench time to 100 C established to be about 2s. The rolling temperature was 500 C and the strain $4s^{-1}$ compared to the extrusion processing conditions of 450 C and $3.83s^{-1}$. The billet was sectioned after quenching and the long transverse plane investigated using polarized light microscopy.

The surface of the specimen was observed to be fully recrystallised, exhibiting a range of grain sizes from 5 to 30μm. In the central (midthickness) regions of the billet, deformed original grains remain (Fig.12a) but with a considerable density of small recrystallised grains (5-10μm) located at the original grain boundaries in addition to those within the grain interiors. The structural variation from the surface to the centre of the billet is substantial and is probably a result of strain variations which are considerable in single pass rolling with the geometry obtaining in this work. The short time interval before quenching almost ensures that the new grains observed could not have been formed by a purely static process and must have nucleated dynamically. It is noteworthy perhaps, that dynamic recrystallisaton has been reported (2) to be a strain dependent phenomenon which is normally observed only if the unit strain is high (e.g. during extrusion). From observation of the hot rolled structure (Fig.12a) the average strain in the central

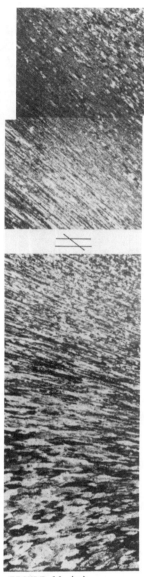

FIGURE 11 (c).
Structural variations
from die face to back
into partially
extruded billet taken
from location 3 of
Fig.1 : discontinuity
indicated.

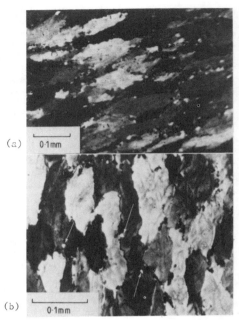

FIGURE 12. Microstructure of mid-thickness
region of (a) hot rolled and (b) cold
rolled and annealed specimen

FIGURE 13. Micrograph of sample structure
taken from (a) homogenized material;
(b) position A at breakthrough pressure.

regions appears to be about 0.25 and hence it can be concluded that the additions of higher solute (Mg) can considerably modify the deformation mode even at low strains.

In order to establish the order of the kinetics of static recrystallisation in the 7%Mg alloy, some material was cold rolled to 40% reduction in thickness and small specimens ($\sim$ 3mm thick) and 1 cm^2 cross section) were annealed in a salt bath for 5s at 475°C (note that quenching after hot deformation occurred well within this time scale). A micrograph of an annealed specimen after quenching is shown in Fig.12b. Recrystallised grains were not observed but a few nuclei (arrowed) could be observed. The scale of recrystallisation, although rapid, is orders of magnitude less than that observed in the hot rolled specimen. Since the dislocation density in such cold rolled material will be two orders of magnitude greater than in the hot rolled material, it is possible to reach the preliminary conclusion that most of the recrystallised grains observed in Fig.11a were nucleated by a dynamic process.

Substructural Investigation

The structure of the alloy after casting and homogenization is shown in Fig.13a. The micrograph indicates that the alloy contains a large density of precipitates but there are also areas which are almost completely particle free. These areas represent regions where dislocation motion and boundary migration will be relatively easy but, in general, the areas are surrounded by dense regions of needlelike precipitates which could act as insurmountable obstacles to further dislocation motion. The development of subgrain structure was similar to that observed in the AA 5456 alloy. Well defined sub-boundaries could be observed, indicating that the Al-7Mg alloy is typical of other Al alloys in that it possesses a high stacking fault energy, thus providing conditions of easy cross-slip and climb to aid the surmounting of obstacles. The micrographs extracted from the billets arrested at location 3 of Fig.1 indicated that very steep deformation gradients exist in the billet and that deformation (related through subgrain size distribution) is much less homogeneous than that observed in the AA 5456 alloy at comparable locations in the extrusion stroke. The subgrains observed were not very well formed and contained large internal dislocation densities (Fig.13b). Micrographs taken from location 4 made no contribution to the definition of deformation mode, being similar, in general, to those from location 3.

At location 5 of Fig.1 considerable deformation was observed throughout the billet. Well developed subgrains and recrystallised grains existed in combination. A large number of recrystallised grains could bne detected coexisting with the dynamically recovered structure observed at position C. Some of these grains were fine (3-8μm) and contained substructures. Typical examples are shown in Fig.14a-c. Fig.14a is an illustration of a recrystallised grain containing a substructure which is just beginning to form subgrain walls. Dislocation clustering can be seen in those parts of the new grain containing fine precipitates. Part of the grain is both precipitate and dislocation free indicating that dislocations tend to move away readily to either an existing or a forming boundary. The selected area diffraction patterns (SADP's) taken across the boundaries of grain A indicate a misorientation of about 45°. Since

169

the grain is only about 5μm in diameter (compared to the original grain size of 115μm), and since no original grain boundaries were observed in the vicinity of this grain, Fig.14a provides sufficient proof that dynamic recrystallisation had occurred. A larger dynamically recrystallised grain containing a few pinned dislocations and some ill defined sub-boundaries is sho3wn in Fig.14b. The most noteworthy feature is that the recrystallised grain is almost particle free while at its boundary there is a precipitate rich region which appears to arrest boundary migration. Another interesting feature is the cluster of larger second-phase particles (arrowed) which lie within the recrystallised grain and may well have provided the nucleation site. The misorientation across the boundaries of the recrystallised grain was found to be about 35°. Figure 14c shows a recrystallised grain located in a precipitate free region. The grain is larger than those observed to contain a substructure and it is likely that it was formed during extrusion and that dislocation annihilation and grain growth occurred during the interruption time before quenching. The structure at positions A and B of Fig.10c, which are closer to the die, was similar to that observed at position C, consisting of subgrains coexisting with dynamic and statically recrystallised grains, though at position A direct evidence of dynamic recrystallisation was less convincing. It is interesting to note that the deformation at position A is less severe than that at either position B or C.

Thus investigation of the partially extruded billets produced substantial evidence that dynamic recrystallisation is an important deformation mode in the Al-7Mg alloy. However, to substantiate the extrusion results, some material was hot rolled in order to reduce any post-deformation activity and foils from longitudinal sections were investigated by transmission electron microscopy.

The microstructure of the hot rolled specimens contained considerably more well formed and larger subgrains than the extruded specimens, which is consistent with the higher temperature employed. The total strain employed was also low ($\dot{\epsilon}$ = 0.69) and, in general, the severe strain gradients which occur during extrusion are not present in the rolling operation. Thus, rolling conditions enabled recovery to take place much more easily than the conditions encountered during extrusion. Nevertheless, numerous recrystallised grains containing a well ordered dislocation substructure were observed. Fig.15a presents an example of a small dynamically recrystallised grain showing dislocations forming into a substructure but also forming a more dense tangle in the centre of the grain where submicrometre size particles appear to be aiding the tangling process by pinning the dislocations. The micrograph shows that the misorientation (obtained from the SADP's) across the boundaries of the recrystallised grain is quite high (30°-37°). The growth of the recrystallised grain has clearly been halted by the quenching operation and also by the barrier of larger particles which are present at the lower boundary of the grain. In addition, it is interesting to note that, at the centre of the recrystallised grain, a 'well-knit' sub-boundary exists, which is typical of a high SFE material. The observation of small size grains, the fairly large dislocation densities, and the presence of well-knit sub-boundaries within the recrystallised grains prove conclusively that dynamic recrystallisation occurs in the Al-7Mg alloy and support the deformation mechanisms suggested above for the extrusion mode.

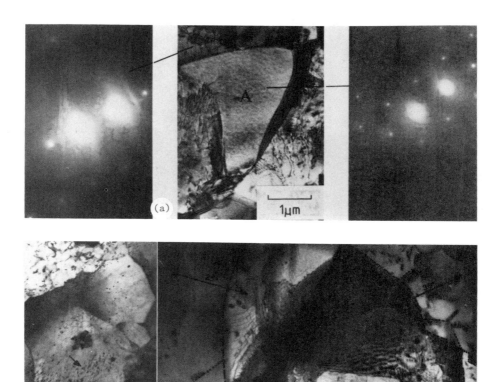

FIGURE 14 (a) (b) (c) Dynamically recrystallized grains at position C in partially extruded billet taken from location 3 (see Figs. 1 and 2(c); in (c) micrograph indicates misorientation across boundaries of dynamically recrytallized grain, x 30 000; (d) recrytallized grain in particle free region showing possible grain growth before quenching.

171

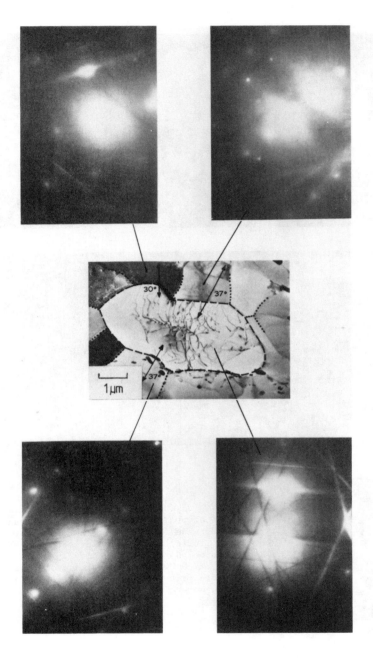

FIGURE 15. Dynamically recrystallized grain in hot rolled specimen showing more uniform distribution of precipitates; misorientation across boundaries of recrystallized grain is indicated.

There were considerable differences in both micro and macrostructural development processes during the extrusion of AA 5456 and Al-7Mg alloy, the structure being much more heterogeneous in the 7%Mg alloy :

(a) In the 5%Mg alloy, samples extracted from the steady state region showed the existence of subgrains at all locations but in the 7%Mg alloy regions consisting of only recrystallised grains were observed.

(b) In the 5%Mg alloy, dynamic recrystallisation was observed only adjacent to the original grain boundaries or at large second phase particles. However, in the 7%Mg alloy, nucleation of dynamic recrystallisation was observed both at these locations and within the grains (e.g. at subgrain boundaries).

(c) For identical deformation and quenching times, recrystallised grains in the 5%Mg alloy were observed only in regions of severe strain (around the die orifice), but in the 7%Mg alloy recrystallised grains were observed at relatively lower strains (even up to 25mm from the die face). In addition, dynamically recrystallised grains were observed (in the 7%Mg alloy) even when the total strain was 0.69).

(d) The structure of the extruded 5%Mg alloy was only partially recrystallised (only in the surface layers), but the 7%Mg alloy extrudate was almost completely recrystallised for similar deformation conditions and quenching times.

Microstructural Observations (Al-10Mg Alloy)

A typical example of the microstructure observed by TEM in the longitudinal sections of all extrudates is presented in Fig.16. Fine equiaxed grains contained within fine 'stringers' of oxide particles are the primary feature of the microstructure. The oxide, which originated as an envelope surrounding the powder particles, is fragmented to form irregular platelets during the extrusion process. There was no evidence of microvoid formation which indicates that there is good contact between the oxide phase and the matrix.

The extrusion temperature had a pronounced effect on the microstructure observed. The TEM shown in Fig16b is taken from the longitudinal section of a rod extruded at 300 C and at a mean strain rate of $4.5s^{-1}$. Diffraction analysis established that the structural elements A, B and C were grains and not subgrains. The absence of fibrous elongated grains and the observation of equiaxed grains containing a substructure in the longitudinal section indicate that those grains are not the result of static recrystallisation after extrusion. Fig.16c is a high magnification micrograph of the grain A of Fig.16b in which dislocation tangles, as well as a well formed sub-boundary, can be identified. A high magnification micrograph of a specimen extruded at 300 C is presented with associated SADP's in Fig.16d. A grain containing some dislocation tangles can be identified clearly, and the SADP's indicate large changes in orientation across the boundary. Since the grain is equiaxed, it must have formed by a process of dynamic recrystallisation very close to the exit of the extrusion chamber. The structure shown by TEM in Fig.16e is of particular interest. A well developed substructure of high dislocation density can be observed. Within the high angle

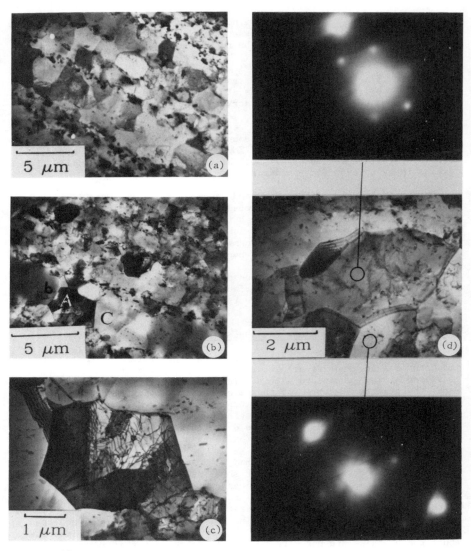

FIGURE 16.

(a) Microstructure typical of that observed in longitudinal sections
of all extrudates;
(b) Microstructure in rod extruded at 300°C (longitudinal section);
(c) Grain A of Fig.16b: high magnification shows existence of well
developed substructure within equiaxed grain;
(d) High magnification micrograph of equiaxed grain containing
dislocation tangles : SADP's indicate complete change in orientation
across boundaries (longitudinal section of specimen extruded at 300°C).

ɒoundaries (marked A) are subgrain walls (marked Ɐ and C). A misorientation of ∿8° is indicated in the SADP's taken across the sub-boundary B. The grain has clearly dynamically recrystallised and this has been followed immediately by the incomplete dynamic recovery in which dislocation tangles have been produced by the application of strain immediately following the recrystallisation process. The evidence suggests that on further straining these tangles form subgrain walls of increasing misorientation: the dislocations multiplying with strain thereby producing boundaries having sufficient misorientation to act as recrystallisation nuclei and thus promoting a further cycle of recrystallisation. Boundary B is quite clearly a probable nucleation site. Dynamically recrystallised grains constrained within oxide stringers were observed for all the extrusion conditions investigated.

Effect of extrusion parameters on dynamically recrystallised grain size (Al-10Mg alloy)

Grain size measurements taken from the TEM's revealed that the recrystallised grain size was dependent on the extrusion temperature. The grain diameter measurements were made from original particles of similar size in order to eliminate the effect of the oxide spacings of about 6μm. During hot deformation, temperature and strain rate are the controlling variables and it was observed that the grain size-temperature relationship was not linear. Therefore, the temperature compensated strain rate Z was plotted against recrsytallised grain size in an attempt to obtain a linear correlation. Luton and Sellars (18) have found that the size of dynamically recrystallised grains found during deformation of ingot stock decrease as Z increases, thus raising the nucleation frequence. Fig.17 shows the relationship obtained, from which it is clear that the grain diameter D increases as Z decreases leading to the linear relationship

$$D^{-1} = 2.02 + 0.08 \ln Z$$

It should be noted from Fig.17 that the grain size is extremely fine, varying from 1.25 to 2.8μm.

Variation of Room Temperature Properties (AA 5456)

The effect on the room-temperature properties produced by varying process parameters was not consistent with results obtained from previous work on other Al-alloy systems in which the tensile strength of the material is related to the subgrain size and hence the temperature compensated strain rate. Figs. 18a and b show the variations produced in proof stress (PS) and UTS.

The results are shown coded by extrusion ratio because, contrary to expectations, it was not possible to identify a single function which could describe a possible relationship. For the lower extrusion ratios, both PS and UTS increase with increasing ln Z but do not exhibit any linear dependence, and the strength decreases at lower temperature. If the structure consisted of subgrains only a linear increase in strength with increasing ln Z would have been expected, as the strengthening effect of smaller, less perfect, subgrains became apparent.

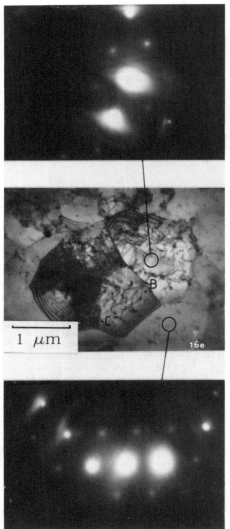

FIGURE 16 (e) Medium angle (8°)
boundary B within dynamically
recrystallized grain
(longitudinal section of
specimen extruded at 300°C).

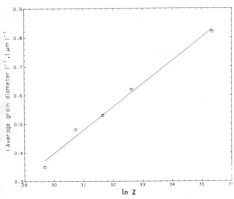

FIGURE 17. Effect of extrusion
parameters on dynamically
recrystallized grain size :
$D^{-1} = 0.08 \ln Z - 2.02$

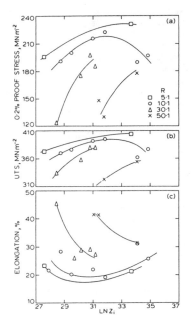

FIGURE 18. Effect of process
variables on extrudate
properties at room
temperature.

The rise in strength may be attributed to both of the mechanisms operating during deformation. The subgrains in the duplex structure will be decreasing in size and, hence, would be expected to strengthen the structure. Concurrent with this subgrain strengthening the dynamic recrystallisation process produces small equiaxed recrsytallised grains of high misorientation in the product. The ease of slip propagation is thus reduced by this high misorientation, hence strengthening the product. This mechanism would suggest, however, that the tensile properties should increase with decreasing Z (i.e. higher temperatures tend to produce greater dynamic recrystallisation) which Fig.18 indicates is clearly not the case. It therefore becomes clear that the growth of dynamically recrystallised grains following processing, andtheir original size, are important structural considerations. Moreover, there will be a driving force in the form of high-angle boundaries which could promote static recrystallisation following deformation which is more likely to occur at higher temperatures. The observation that tensile properties are dependent on total strain, as indicated in Fig.18, would also support this latter view. It is well known that during torsion testing (of copper and steels) a critical strain is required to induce dynamic recrystallisation, and this strain is about 0.75 times the peak strain. Torsion testing of this alloy (11) revealed the maximum possible strain for the onset of dynamic recrystallisation to be 0.35. while the total strain corresponding to the lowest extrusion ratio (R = 5) is 1.61. Experimental observations also indicated that some dynamic recrystallisation could be observed in all the extrusion experiments performed. Nevertheless, higher extrusion ratios will promote dynamic recrystallisation close to the entry of the deformation zone and there will thus be a propensity for a greater degree of dynamic recrystallisation in these extrudates performed at higher extrusion ratios. The propensity for subsequent static recrystallisation will also be increased. The decrease in properties can only be associated with the gradual cessation of dynamic recrystallisation during working. Clearly,the presence of the small dynamically recrystallised grains within the subgrain matrix enhances the strength, suggesting that this grain-strengthening effect is much greater than subgrain strengthening. The results also indicate that the volume fraction of dynamically recrystallised grains present in the material was considerably greater than indicated by the electron-microscopy results. This is not surprising, since a proportion of such grains will, subsequently, undergo a dynamic recovery process. The results shown for extrusion ratios of 30:1 and 50:1 do not reach peak values because it was not possible to extrude at a high enough Z value to prevent dynamic recrystallisation.

Conclusions

Dynamic recrystallisation has been observed in the three Al-Mg alloys investigated in this work. The 5% and 7% Mg alloys displayed a duplex deformation mode which included dynamic recovery. The incidence of dynamic recrystallisation was greater in the higher magnesium containing alloy. This is not related to the action of Mg atoms in binding dislocation motion by decreasing the density of active vacancies. Nevertheless second phase particles are still required to nucleate recrystallisation. In the 10%Mg alloy dynamic recrystallisation was complete probably due to the restraining effect on recovery of oxide stringers and the subsequent larger fraction of particles available for nucleation.

Acknowledgements

    The authors would wish to express their appreciation of continued
support from ALCAN INTERNATIONAL who supplied the materials used in
this work and considerable financial backing. Student support has also
been supplied by the Science and Engineering Research Council and this
is also acknowledged.

References

1.   H. J. McQueen, J. Met. 1980, 32, 17.

2.   H.J. McQueen and J.J. Jonas : Treatise on Materials Science
     and Technology, Vol.6, pp.393-493, 1975, New York Academic Press.

3.   S. Dulop and H.J.McQueen : Superalloy Processing HI-H21, 1972,
     Ohio Metals and Ceramics Information Center.

4.   C. Rossard, Rev. Metall., 1968, 65, 181.

5.   J.N. Greenwood and H.K. Warner, J.I.M., 1969, 64, 135.

6.   D. Hardwick and W.J. McG. Tegart, Mem.Sci.Rev.Metall., 1961, 58, 869.

7.   W. Truszkowski, A. Pawlowski and J. Dutriewicz, Bull.Acad.Pol.Sci.,
     1971, 19, 55.

8.   H.M. Tensi, P. Dropman and H. Borches, Z. Metallkd. 1970, 61, 518.

9.   K.J. Gardiner and R. Grimes, Met.Sci. 1979, 13, 216.

10. M.A. Zaidi, Ph.D. Thesis, University of London, 1980.

11. M.G. Tutcher, Ph.D. Thesis, University of London, 1979.

12. H.J. McQueen, E. Evangelista, J. Bowles and G. Crawford, Met.Sci.

13. T. Sheppard and M.G. Tutcher, Met.Sci., 1980, 14, 579.

14. T. Sheppard, N.C. Parson and M.A. Zaidi, Met.Sci., 1983, 17, 481.

15. T. Sheppard, et al - this conference.

16. M.G. Tutcher and T. Sheppard, Met.Technol. 1980, 7, 488.

17. T. Sheppard, M.G. Tutcher and H.M. Flower, Met.Sci., 1979, 13, 473.

18. M.J. Luton and C.M. Sellars, Acta Metall. 1969, 17, 1033.

RECRYSTALLIZATION CHARACTERISTICS OF ALUMINIUM - 1% MAGNESIUM

UNDER HOT WORKING CONDITIONS

C. M. Sellars*, A. M. Irisarri** and E. S. Pucni***

* Department of Metallurgy, University of Sheffield,
Sheffield S1 3JD, England.
** Now at INASMET, San Sebastian, Spain.
*** Now at Facultad de Ingenieria, Universidad
Central de Venezuela, Caracas, Venezuela.

Summary

Plane strain compression tests have been carried out using interrupted deformation to determine the restoration index as a function of time of holding between deformations. Specimens have also been quenched and examined metallographically to determine the correlation between restoration index and recrystallized fraction, and to measure recrystallized grain sizes. The ranges of conditions examined were: temperature 510-350°C, strain in the first deformation 0.15-1.5, strain rate 0.1-3.76 s^{-1}, initial grain size 54-112 μm. From the results, quantitative relationships describing recrystallization kinetics and recrystallized grain size in terms of the processing variables are proposed. These relationships are of suitable form for incorporation in computer models of microstructural evolution during industrial hot rolling operations.

Introduction

Hot working operations are carried out primarily to change the shape of the material from that of the original cast blocks to a wide range of semi-finished and finished products, but the thermomechanical treatments also have a major effect on the microstructure. Operations such as rolling impose the deformation in a series of passes separated by intervals of time. The final microstructure is therefore determined by the interaction between the dynamic structural changes, which take place during deformation, and the static structural changes, which take place between the deformation passes and after the deformation processing is complete. The process resulting in the most obvious changes in microstructure is static recrystallization, which replaces the deformed grain structure by new strain-free grains. Whether this is important in a specific working operation depends on the kinetics of static recrystallization in relation to the times available during processing.

The kinetics of static recrystallization are strongly influenced by alloy composition, homogenisation treatment and thermomechanical processing conditions. In a recent survey (1) of the influence of process variables on recrystallization rate and recrystallized grain size, it was shown that strain (ε), strain rate ($\dot{\varepsilon}$) and temperature of deformation (T_{def}), and temperature of holding (T) are each important variables and so is the grain size (d_0) before deformation. Relationships were proposed, but the paucity of quantitative data for aluminium and aluminium alloys made these relationships extremely tentative. The purpose of the present work is to provide quantitative data on the effects of each of the above variables for one aluminium alloy and to propose more soundly based relationships.

Experimental Materials and Procedure

The experimental work was carried out on two casts of aluminium - 1% magnesium alloy of analyses shown in Table I. The materials were produced

Table I. Analysis of Experimental Alloys, wt.%

Alloy	Cu	Fe	Mg	Mn	Si	Ti	Zn
A	0.16	0.40	0.95	< 0.02	0.10	0.014	< 0.05
B	0.16	0.47	1.00	0.005	0.12	0.013	0.02

by DC casting laboratory melts to 26 mm thick slab. Material A was supplied[*] as 10 mm thick hot rolled plates, prepared by homogenising the slab for 16 hours at 550-560°C and rolling directly after homogenisation, with finishing temperatures of 250-300°C. Material B was supplied[*] as cast slab. This was given a nominally identical homogenisation treatment and then air cooled to room temperature. Some specimens were machined from the slab in this condition, but the majority of the material was reheated to 400°C, rolled in three passes to 10 mm plate and quenched. In a few cases the reheating temperature was varied to obtain different grain sizes in the test specimens.

---

[*] Alcan International Laboratories, Banbury.

Plane Strain Compression Testing

Specimens for plane strain compression testing were cut directly from the rolled plates. The specimens were 80 mm long x 50 mm wide with the long side parallel to the rolling direction. All rolled specimens were annealed at 525°C for 1 hour and air cooled to ensure that they were in a fully recrystallized condition before reheating for testing. In order to determine the centre temperatures of specimens during deformation, 1.5 mm dia. holes were drilled from the centre of one long edge to the mid-plane and Pyrotenax mineral insulated chromel-alumel thermocouples were inserted. All specimens were coated with a graphite-molybdenum disulphide lubricant over the area to be contacted by the plane strain compression tools.

Plane strain compression tests were performed on the computer controlled servohydraulic machine described elsewhere (2,3). The compression tools (13%Cr hot die steel) have 15 mm wide x 100 mm long working faces which deform the specimens centrally across their width direction. The machine is provided with a reheating furnace and a separate furnace which surrounds the tools. In the early stage of this work it was found that because of heat conduction down the tools,the working faces were at a lower temperature than the environmental temperature within the furnace. The difference in face temperature was always greater for the bottom tool than for the top tool; the values rising to a maximum of 80° and 60°C respectively at an environmental temperature of 300°C. The temperature difference caused some chilling of the specimens during deformation, followed by a return to the set environmental temperature within 10s after deformation when the specimens were held within the furnace, but out of contact with the tools.

In the course of this work the tools were modified by drilling them across the breadth 25 mm from the water cooled ends and inserting 600 watt cartridge heaters. These were separately controlled by controlling the power input through variable transformers to bring the working faces to the same temperature as that set in the test furnace. With the heaters in operation, the deformational heating caused a rise in temperature of the specimens, particularly during tests at higher strain rates. This again returned to the set environmental temperature within 10s after deformation.

For the majority of tests,the restoration kinetics were determined from the change in flow stress at 0.05 strain in a second deformation carried out at a series of predetermined time intervals after the end of the first deformation. Conditions of strain rate and temperature in the second deformation were always the same as those applied in the first deformation and were in the range 0.1 to 3.67 s^{-1} equivalent tensile strain rate and 350 to 510°C. All stresses and strains were converted to the equivalent tensile values using the Levy-von Mises relationships after correction had been made for the effects of friction and lateral spread. A restoration index was determined as:

$$R = \frac{\sigma_1 - \sigma_2}{\sigma_1 - \sigma_2^{\infty}} \qquad (1)$$

where σ_1 is the value of flow stress at the end of the first deformation (extrapolated to a strain of 0.05 in the second deformation) σ_2 is the flow stress at a strain of 0.05 in the second deformation and σ_2^{∞} is the equivalent flow stress measured after a long time interval between the deformations when there was no further softening with increasing time.

181

Metallography

For comparison with the restoration kinetics, recrystallization kinetics were measured on some specimens by giving them the first deformation followed by a series of holding times at temperature and then quenching them. These specimens were sectioned longitudinally, at mid-plane. Strips about 10 mm wide were obtained, from which the deformed region was cut, cleaned in hydrochloric acid and mounted in slow resin. A hole was then drilled through the resin so that a stainless steel screw could be inserted to make electrical contact. Specimens were ground and polished on diamond wheels to $\frac{1}{4}$ μm and then electropolished in a solution of 95% ethyl alcohol and 5% perchloric acid at $0^{o}C$ using fast stirring. The electropolishing was carried out using a stainless steel cathode and a potential of 27 volts. Specimens were then anodised in a solution of 46 ml tetrafluoroboric acid and 7 g boric acid in 970 ml water, using a stainless steel cathode with a potential of 20 volts and a maximum current density of $0.2A/cm^2$. The time of anodising had to be increased with the degree of recrystallization to obtain satisfactory results for optical examination in a polarising microscope. To measure fraction recrystallized, point counting was carried out along 16 lines through the deformed thickness. Sixteen lines approximately 1 mm apart were used so that the whole of the deformed width was covered. Point counting was carried out using a calibrated eyepiece with 21 points on the scale at a magnification such that no two adjacent points fell in the same grain. After each count the specimen was moved so that the adjacent area on the traverse line was measured. In this way the distribution of recrystallization in the deformed zone was determined (4), but in this paper only the mean value for each specimen is reported.

The grain size of undeformed specimens and of fully recrystallized specimens was determined as the mean linear intercept

$$\bar{d} = (\bar{L}_T . \bar{L}_L)^{\frac{1}{2}} \tag{2}$$

where $\bar{L}_T$ and $\bar{L}_L$ are the mean linear intercepts measured on lines 1 mm apart through the thickness and in the longitudinal direction (in the deformed zone), respectively.

Results and Interpretation

All the stress-strain curves showed an initial rise in flow stress, leading to steady state at strains of about 0.2 to 0.5 depending on deformation conditions. The typical form is shown by the line for a continuous deformation at $400^{o}C$ and $1 \ s^{-1}$ in figure 1. This figure also shows the effect of interrupting the deformation for different periods of time after an initial strain of either 0.15 or 0.67. For clarity in this figure, flow stresses for each test have been adjusted slightly so that the first deformation coincides exactly with that of the continuous deformation curve. In practice, flow stress in the first deformation varied randomly by about $\pm$ 3%.

Restoration and Recrystallization Kinetics

The restored fraction determined using equation (1) for the curves in figure 1 and equivalent curves for initial strains of 0.33 and 1.00 is shown as a function of holding time in figure 2. This shows rapid apparent restoration during the first 10 s holding period, which increases with increasing initial strain. The initial apparent restoration is attributable

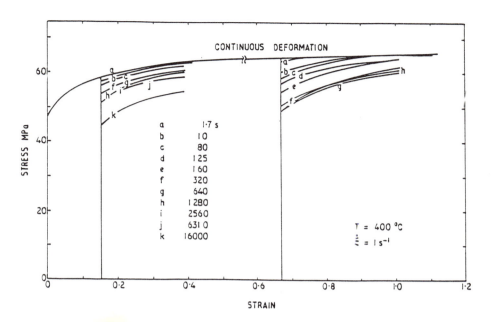

Figure 1 - Equivalent tensile stress-strain curves for Material A given two deformations (without tool heaters) with an initial strain of either 0.15 or 0.67. Initial grain size 90 μm.

directly to the fall in mean temperature during the first deformation because these tests were performed without tool heaters. From thermocouple measurements during deformation the change in centre temperature was measured directly. Such measurements were also used to obtain the heat transfer coefficient of 150 kWm^{-2}K^{-1} between the specimen and tools, which was used in a finite difference computer program (5) to obtain the change in mean temperature of the deformation zone with strain. For the conditions of figure 2 these changes are -11, -13, -16 and -20°C for strains of 0.15, 0.33, 0.67 and 1.00, respectively. From the temperature dependence of flow stress these temperature changes, which are eliminated within 10 s of holding, would lead to apparent restored fractions of more than the observed values if temperature were uniformly distributed during deformation. At times longer than 10 s, the nearly linear increase in restored fraction with logarithm of holding time represents true recovery, with the accelerated restoration arising from recrystallization, as shown by the measurements of recrystallized fraction in figure 3. Recrystallization does not occur homogeneously throughout the deformed zone, as illustrated in figure 4, but takes place preferentially along the bands of high local strain. These have a characteristic pattern related to the expected slip line fields (6) and result in a change in shape of the mean fraction recrystallized versus time curves with increasing strain. This is shown in figure 3 and will be discussed in detail elsewhere. Because final recrystallization takes place in regions away from the active slip line fields, restoration is complete before recrystallization is complete, as shown in figure 5. In this figure the results plotted are the individual ones for separate specimens held for the same times for restoration and for recrystallization measurements. This gives a wide scatter of points, which is reduced if the mean curves in figures 2 and 3 are correlated, but this correlation gives the same trend,

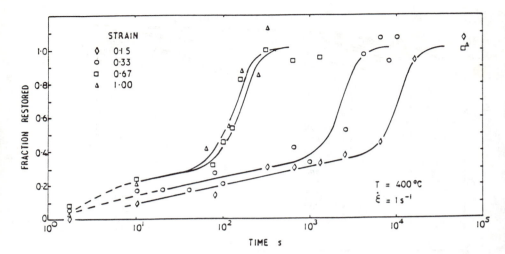

Figure 2 - Dependence of restored fraction on holding time for Material A given different initial strains (without tool heaters). Initial grain size 90 μm

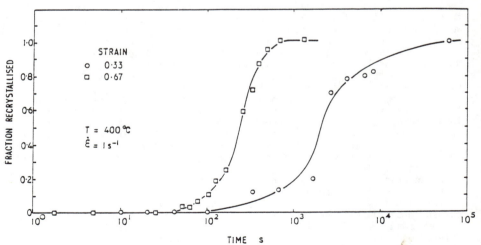

Figure 3 - Dependence of recrystallized fraction in the deformed zone for Material A given initial strains of 0.33 or 0.67 (without tool heaters). Initial grain size 90 μm.

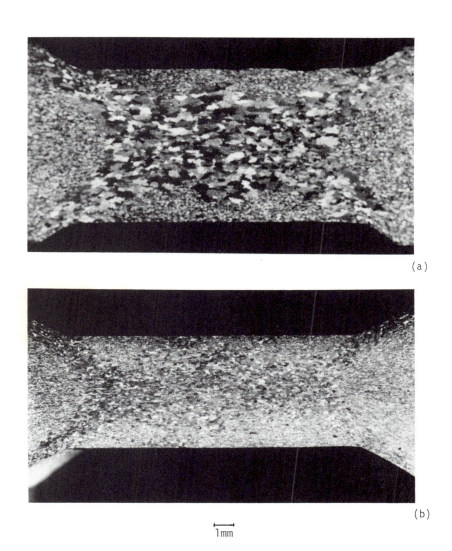

(a)

(b)

1mm

Figure 4 - Macrographs of the deformed zone of specimens of Material A deformed (without tool heaters) at 400°C and 1 s⁻¹ (a) to a strain of 0.33 and held for 2560 s and (b) to a strain of 0.67 and held for 160 s.

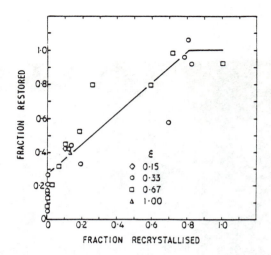

Figure 5 - Relationship between fraction restored and fraction recrystallized in Material A deformed to various strains at 400°C and 1 s⁻¹.

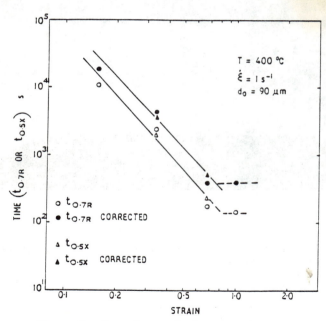

Figure 6 - Dependence of time to 0.7 fraction restored and 0.5 fraction recrystallized on prior strain. Filled points show the values corrected for the change in temperature during deformation without tool heaters.

with 0.5 fraction recrystallized corresponding to about 0.7 fraction restored.

Effect of Strain

The times for these fractions restored or recrystallized are shown as a function of strain in figure 6. The open points are the measured values and the filled ones have been obtained by taking account of the drop in temperature during deformation leading to an increase in Zener Hollomon parameter, in the way described later, to give the times expected for a strain rate of 1 s^{-1} at a constant temperature of 400°C. Clearly, prior strain has a major effect on the kinetics of recrystallization at least up to a strain 0.8. The line through the corrected points gives

$$t_{(0.7R)} \propto \varepsilon^{-2.7} \tag{3}$$

Effect of Strain Rate and Temperature

Restoration curves were determined on Material B after deformation to a strain of 0.67 at different strain rates and temperatures in the range 350 to 510°C. The times to 0.7 fraction restored are shown as a function of reciprocal temperature in figure 7. It can be seen that, apart from the result at 350°C and 1 s^{-1} which took a much longer time than expected, the results at constant strain rate fall on straight lines. These show a large effect of the strain rate of deformation and a relatively small effect of temperature. At a constant temperature the displacement of the lines indicates that

$$t_{(0.7R)} \propto \dot{\varepsilon}^{-1.1} \tag{4}$$

and at constant strain rate the slope leads to an apparent activation energy (Q_{app}) for recrystallization of about 58 kJ/mol. Because the driving force varies with temperature when deformation is carried out at a constant strain rate, the true activation energy for recrystallization (Q_{rex}) can only be obtained when results are compared after deformation at different strain rates for different temperatures such that the Zener Hollomon parameter ($Z = \dot{\varepsilon}$ exp 156000/RT) is constant and hence the flow stress at a given strain and the driving force are constant. On this basis, the relationship of equation (4) was considered to arise from the effect of strain rate on Z, so that

$$t_{(0.7R)} \propto Z^{-1.1} \tag{5}$$

Equation (5) was then used to correct the observed values of $t_{(0.7R)}$ for the temperature rise during deformation (which ranged from C.5°C at 0.1 s^{-1} and 460°C to 6.5^{o}C at 1 s^{-1} and 390°C) so that the filled points in figure 7 represent the times expected for isothermal deformation at the initial (and holding) temperatures. This correction has only a minor effect on the slope of the lines, but enables the locus of points for a constant value of $Z = 9.7 \times 10^{10}$ s^{-1} to be obtained. This is a straight line which gives the value of $Q_{rex} = 230$ kJ/mol. Thus

$$t_{(0.7R)} \propto Z^{-1.1} \exp Q_{rex}/RT \tag{6}$$

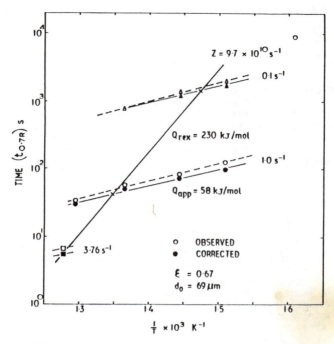

Figure 7 - Dependence of time to 0.7 fraction
restored on temperature for Material B deformed
(with tool heaters) to a strain of 0.67 at various
temperatures. Filled points show the values
corrected for the rise in temperature during
deformation.

Substituting for Z in this equation gives

$$t_{(0.7R)} \propto \dot{\varepsilon}^{-1.1} \exp(-1.1\,Q_{def} + Q_{rex})/RT \tag{7}$$

which means that, when $\dot{\varepsilon}$ is constant, the apparent activation energy for
recrystallization,

$$Q_{app} = -1.1 \times 156 + 230 = 58.4 \text{ kJ/mol} \tag{8}$$

in agreement with the value given earlier.

The importance of the final value of Z during deformation on the
recrystallization kinetics is clearly illustrated in figure 8. This shows
a direct comparison of the restoration curves obtained for Material B tested
to the same strain at a strain rate of 1 s^{-1} with an environmental temperature of 400°C, but without the tool heaters for one series of tests and with
the tool heaters for the other. For the specimens tested without tool
heaters the mean temperature fell to 384°C by the end of deformation and
returned to 400°C within 10 s. From the value of Q_{rex} given earlier, a
correction can be made for the low temperature during the initial period of
holding using the procedure described previously (1). This leads to the
filled points expected for isothermal annealing at 400°C, which are at only

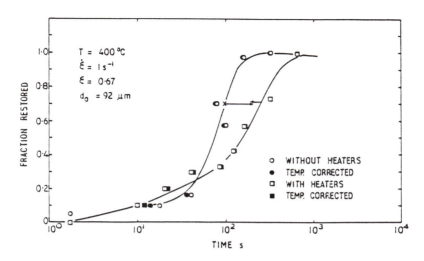

Figure 8 - Comparison of the restoration kinetics of Material
B tested with and without tool heaters at an environmental
temperature of 400°C. Arrows show the shifts expected if
changes in Z during deformation are allowed for.

slightly shorter times than the actual times. With tool heaters, the mean
temperature rose to 406°C by the end of deformation, leading to a shift of
the filled points for isothermal annealing at 400°C to slightly longer times
than the actual times. It can be seen that these corrections do not bring
the two curves into closer coincidence except at short times before
recrystallization starts. However, the final Z values for deformation without and with heaters differ because of the different final temperatures of
deformation and are 2.57×10^{12} and 1.02×10^{12} s^{-1} respectively. Using
equation (5) to correct $t_{(0.7R)}$ to the value of $Z = 1.30 \times 10^{12}$ s^{-1}
expected for isothermal deformation at 400°C leads to the shifts shown by
the arrows. It can be seen that the heads of the arrows virtually coincide,
illustrating clearly the major effect of the final value of Z.

Recrystallized Grain Size

   Recrystallized grains were always slightly elongated in the longitudinal
direction, presumably because of the influence of the second phase particles.
The quantitative measurements of grain size reported are the values of $\bar{d}$
obtained from equation (2), which represent the equivalent equiaxed grain
sizes for plane strain conditions.

   For tests on Material B carried out at the same strain rate and
different temperatures, the grain size in fully recrystallized specimens is
shown as a function of Z in figure 9. The values of Z for the open points
are calculated for the initial values of temperature of deformation whereas
the filled points are calculated from the final, somewhat higher temperatures
attained with the use of tool heaters. The shift is relatively small, but
increases with increase in Z, i.e. with decrease in initial temperature. The
line through the filled points gives a relationship

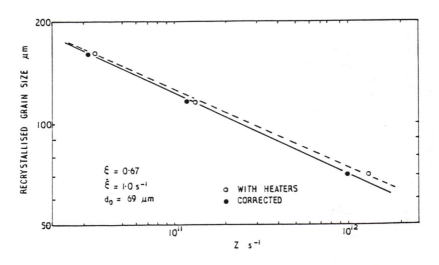

Figure 9 - Dependence of recrystallized grain size in Material B
on Z. The open points show the initial values of Z and the filled
ones the final values of Z.

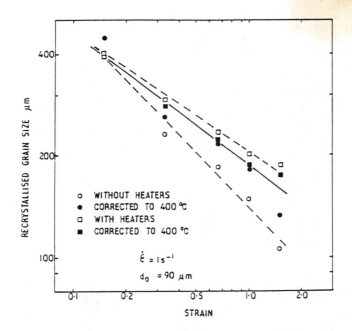

Figure 10 - Dependence of recrystallized grain size
in Material A on Strain. The open points show the
experimental measurements and the filled ones the
values expected for isothermal deformation at 400°C.

$$d_{rex} \propto Z^{-0.24} \tag{9}$$

The effect of strain on recrystallized grain size was determined on Material A from tests carried out both with and without tool heaters. The results obtained are shown by the open points in figure 10. These differ markedly for a given strain for the tests with and without tool heaters. Using equation (9) to correct the values of d_{rex} to those expected for iso-thermal deformation at 400°C leads to the filled points. These are in reasonable coincidence except at a strain of 1.5 and can be reasonably fitted by a straight line giving the relationship for constant Z of

$$d_{rex} \propto \varepsilon^{-0.39} \tag{10}$$

Effect of Initial Grain Size

Different initial grain sizes were obtained in Material B by varying the prior thermomechanical processing. The effect on the restoration curves after deformation under constant conditions of temperature, strain rate and strain is illustrated in figure 11. The data points show considerable scatter, but for the three smaller grain sizes they fall around curves of similar form. For the largest grain size of 112 μm obtained in the as-cast and homogenised slab, the points could be fitted by a curve of similar form, shown solid, or by one of lower slope, shown by the broken line. Shortage of cast material prevented this from being investigated further, but in the analysis of the data, the solid curve has been used. This leads to a time for 0.7 fraction restored which is consistent with the effect of grain size given by the other data, figure 12(a). This seems to be reasonable in view of the consistency of the results for recrystallized grain size shown in figure 12(b). The straight lines drawn in these figures give, respectively,

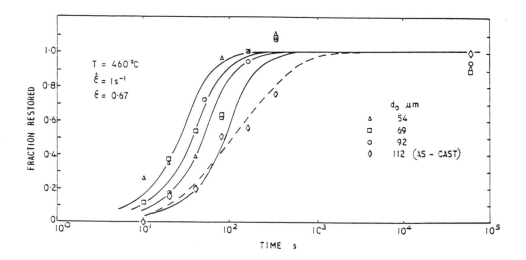

Figure 11 - Effect of initial grain size on the restoration curves for Material B tested at 1 s⁻¹ and 460°C with tool heaters.

191

the relationships

$$t_{(0.7R)} \propto d_o^{1.35} \qquad (11)$$

and
$$d_{rex} \propto d_o^{1.3} \qquad (12)$$

Discussion

If each of the variables studied is considered to act independently,

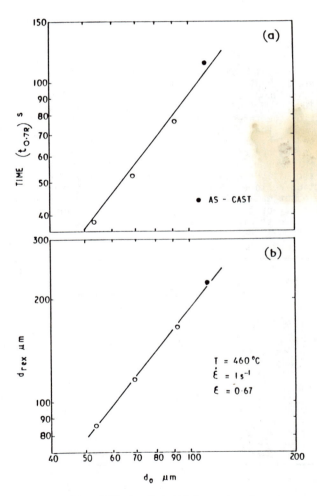

Figure 12 - Effect of initial grain size on
(a) time for 0.7 fraction restored and
(b) recrystallized grain size for Material B
tested at 1 s^{-1} and 460°C with tool heaters.

192

the separate equations (3), (6) and (11) for $t_{(0.7R)} \simeq t_{(0.5X)}$ can be combined to give the overall relationship

$$t_{(0.5X)} = \alpha d_o^{1.35} \, \varepsilon^{-2.7} \, Z^{-1.1} \, \exp 230000/RT \tag{13}$$

where T is the temperature of holding and Z is the value of Zener Hollomon parameter at the end of the deformation. Similarly, from equations (9), (10) and (12)

$$d_{rex} = \beta d_o^{1.3} \, \varepsilon^{-0.39} \, Z^{-0.24} \tag{14}$$

α and β in equations (13) and (14) are material constants that depend on composition and homogenising treatment. The two materials used in this investigation were similar in composition and had been given similar homogenisation treatments. However, the constant α was found to have mean values of 9.8×10^{-6} and 3.8×10^{-6} $\mu m^{-1.35}$ $s^{-0.1}$ and the constant β values of 4.35×10^2 and 1.85×10^2 $\mu m^{-0.3}$ $s^{-0.24}$ for Materials A and B respectively. These differences are considered to arise from relatively small differences in distribution of second phase particles in the alloys caused by slight variations in the casting conditions or homogenisation treatments. Zaidi and Sheppard (7) have shown that for aluminium-magnesium alloys 550°C is a critical temperature for homogenisation; a variation of ± 10°C producing a change in recrystallization temperature approaching 100°C for identical deformation conditions in alloy AA5454. From equation (13) the factor of 2.35 in the values of α for Materials A and B implies a difference in recrystallization temperature of only about 15°C.

Using the values of α given above, the observed restoration times are compared with the values calculated from equation (13) for strains < 0.67 in figure 13. It can be seen that the results fall close to the line of unit slope indicating that equation (13) gives satisfactory correlation of the experimental data. Figure 6 indicates that the strain dependence may change at strains above about 0.8 for deformation at 400°C and 1 s^{-1}. Such a change would be consistent with the change in strain dependence found in α-iron at the onset of steady state produced by work hardening and dynamic recovery (8). The present study did not cover sufficiently high strains to determine whether $t_{(0.7R)}$ becomes constant or continues to decrease with strain in the steady state region. However, previous analysis of data for aluminium alloys for strains > 0.59 indicates that a strain dependence with an exponent of about -1 may extend to strains well into steady state (1). It is also noteworthy in figure 7 that restoration at 350°C requires a much longer time than predicted by equation (13). This is thought to occur because of secondary precipitation, either during reheating for testing or during holding after testing, when the temperature is much below that of homogenisation.

In a similar manner to figure 13, figure 14 shows that equation (14) satisfactorily correlates the data for recrystallized grain size. This is true for all the data for strains up to and including 1.5. The observation that the relationship between recrystallized grain size and strain does not change until considerably higher strains than those to the onset of steady state deformation is again consistent with results on α-iron (3).

Comparison of equations (13) and (14) with the equivalent ones for steels in the austenitic condition (9) shows that strain has a somewhat smaller effect on both recrystallization kinetics and recrystallized grain size, probably reflecting the lower work hardening rate of Al-1%Mg resulting

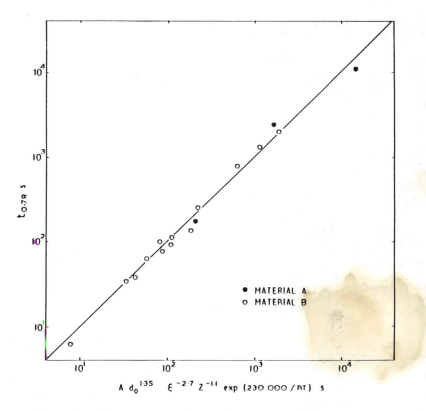

$A \ d_0^{1\cdot35} \quad \dot{\varepsilon}^{-2\cdot7} \ Z^{-1\cdot1} \ exp \ (230 \ 000 \ / \ RT) \quad s$

Figure 13 - Comparison between observed and predicted times for 0.7 fraction restored (0.5 fraction recrystallized) for all conditions of deformation at temperatures of 390 to 510°C and strain rates of 0.1 to 3.76 s^{-1} to strains of $\leqslant$ 0.67.

from more rapid dynamic recovery. Original grain size has a smaller effect on recrystallization kinetics but a larger effect on recrystallized grain size than in steels. The reason for this is unclear, but it is noteworthy that, whereas most hot working conditions lead to grain refinement of austenite, the deformation conditions of the present study usually give grain coarsening by recrystallization. The most striking difference between the equations for steels and for this aluminium alloy is in the influence of Z, which is small for steels but large in both equations (13) and (14). This large effect of Z may arise from the establishment of sub-grain structures, whose size varies inversely with Z (10), from relatively low strains in aluminium alloys. As discussed earlier the value of Z of importance appears to be the one at the end of deformation when Z changes only relatively slowly, as in the present experiments. This implies that the microstructure during deformation changes in step with the change in Z, a conclusion that has been substantiated by more recent work (11).

Although more study is required before definitive interpretations of equations (13) and (14) can be made, they do provide an empirical basis

194

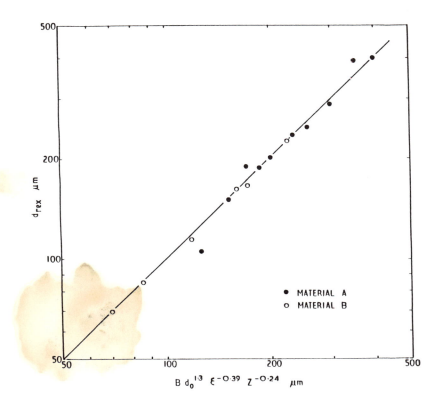

Figure 14 - Comparison between observed and predicted recrystallized grain sizes for deformation at temperatures of 390 to
510°C to strains of ⩽ 1.5.

from which the behaviour of aluminium - 1% magnesium during industrial
rolling can be modeled in the same way as has been done previously for
steels (9) to obtain quantitative predictions of the microstructural changes
in terms of the variables of plate or strip rolling. The results of such
modeling (4) will be reported elsewhere, but they confirm the previous
conclusions from modeling of hot working of steels that combining limited
quantitative laboratory data with computer models provides a powerful method
for understanding the complex interactions of material and process variables
that take place during industrial thermomechanical processing.

Conclusions

From studies of the kinetics of restoration and recrystallization and
from measurements of recrystallized grain size of aluminium - 1% magnesium
alloys after high temperature deformation in plane strain compression, the
effect of deformation variables and of initial grain size on recrystallization time and recrystallized grain size can be satisfactorily described by
equations in which each of the variables is considered to act independently.
The material constant in these equations is sensitive to the prior homogeni-

sation conditions. The equations provide a quantitative basis for computer modeling the sequential structural changes that take place during industrial hot rolling operations of this alloy.

Acknowledgements

The authors are grateful to the Gobierno Vasco, Spain for the provision of a scholarship for AMI and to La Facultad de Ingenieria de la Universidad Central de Venezuela and El Consejo de Desarrollo Cientifico y Humanistico for financial support of ESP.

References

1. C. M. Sellars, "Hot Working Operations", pp. 405-440 in Aluminium Transformation Technology, C.A. Pampillo, H. Biloni and D. E. Embury, eds.; ASM, Metals Park, Ohio, 1980.

2. C. M. Sellars, J. P. Sah, J. Beynon and S. R. Foster, "Plane Strain Compression Testing at Elevated Temperatures", Dept. of Metallurgy Report, University of Sheffield, 1976.

3. J. P. Sah and C. M. Sellars, "Effect of Deformation History on Static Recrystallization and Restoration in Ferritic Stainless Steel", pp. 62-66 in Hot Working and Forming Processes, C. M. Sellars and G. J. Davies, eds.; The Metals Society, London, 1980.

4. E. S. Puchi Cabrera, "Effect of the Deformation History on the Recrystallization Kinetics of Al-1%Mg Alloy", Ph.D. Thesis, University of Sheffield, 1983.

5. S. R. Foster, "Simulation of Hot Rolling of Low Carbon Steels", Ph.D. Thesis, University of Sheffield, 1981.

6. John H. Beynon and C. Michael Sellars, "Strain Distribution Patterns during Plane Strain Compression", J. Testing and Evaluation, 13(1), (1985), pp. 28-38.

7. M. A. Zaidi and T. Sheppard, "Effect of High Temperature Soak and Cooling Rate on Recrystallization Behaviour of Two Al-Mg Alloys (AA 5252 and AA 5454)", Metals Technology, 11(8), (1984), pp. 313-319.

8. G. Glover and C. M. Sellars, "Static Recrystallization after Hot Deformation of α-Iron", Metallurgical Transactions, 3, (1972), pp. 2271-2280.

9. C. M. Sellars, "The Physical Metallurgy of Hot Working", pp. 3-15 in Hot Working and Forming Processes, C. M. Sellars and G. J. Davies, eds,; The Metals Society, London, 1980.

10. H. J. McQueen and J. J. Jonas, "Recovery and Recrystallization during High Temperature Deformation", pp. 393-493 in Plastic Deformation of Materials, R. J. Arsenault, ed.; Academic Press, New York, (1975).

11. F. R. Castro-Fernandez, "Microstructural Changes during Multi-Pass Hot Working of an Al-1%Mg-1%Mn Alloy", Ph.D. Thesis, University of Sheffield, 1985.

RECOVERY AND RECRYSTALIZATION IN THE

HOT-WORKING OF ALUMINUM ALLOYS

H.J. McQueen and K. Conrod

Mechanical Engineering Department
Concordia University
Montreal, Quebec
Canada   H3G 1M8

Summary

The steady state or peak flow stresses of Al, Al-Mg and Al-Mg-Mn alloys are related to the temperature by an Arrhenius function and to the strain rate by the hyperbolic sine function. The effect of Mg and impurities on the above results is examined. The mechanism of dynamic recovery is analyzed through the development of grain and subgrain structures in Al deformed at 400°C, 0.2 s^{-1} through strains between 1 and 60. The mecharism of dynamic recrystallization due to nucleation enhancement at large particles in 5083 alloy was studied by optical and transmission microscopy. Static recovery and recrystallization after hot working were determined in Al by SEM channeling contrast and in Al-Mg alloys by optical and mechanical metallography and the kinetic parameters calculated.

Introduction

The objective of this review is a synthesis of the significant results on Al and its alloys recently published by the authors' colleagues along with some pertinent work from the literature. The emphasis will be placed on the mechanisms of dynamic and static, recovery and recrystallization and on their effect on temperature and rate dependence of the flow stress [1-10]. The influences on the restoration mechanisms of solute and dispersed particles as found in Al-Mg and Al-Mg-Mn alloys are examined but age hardenable alloys are not considered [11-12].

The research reviewed commences chronologically with a determination of static recovery during a 200 h, 200°C anneal of specimens hot compressed at 0.1 - 200 s^{-1}, 25-500°C by means of channeling mode SEM [13-14]. The flow stress in torsion of Al-4.5Mg-0.8Mn was determined over the range 300-500°C, 0.1-1.0 s^{-1}. The dynamically recovered and recrystallized grain and substructure at ε = 5 were examined by optical microscopy and TEM [15-17]. Multistage tests were performed to determine static softening between passes. The flow stresses in torsion of high purity Al and Al-Mg alloys were determined at 0.4 - 0.8 s^{-1} from 25-600°C and compared to previous research results [18]. The static recrystallization after strains up to 2.45 was determined from held and quenched samples by optical microscopy and also by means of duplex mechanical testing. The development of grair and subgrain structures in commercial Al was studied over strains of 1, 3, 10, 20, 40 and 60 by means of polarized light microscopy, SEM channeling contrast and TEM [19-20]. These results are presented in the order given in the abstract.

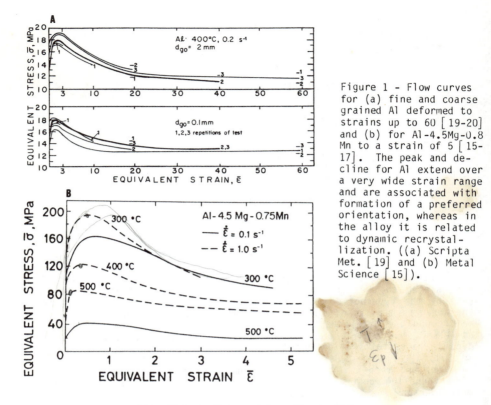

Figure 1 - Flow curves for (a) fine and coarse grained Al deformed to strains up to 60 [19-20] and (b) for Al-4.5Mg-0.8 Mn to a strain of 5 [15-17]. The peak and decline for Al extend over a very wide strain range and are associated with formation of a preferred orientation, whereas in the alloy it is related to dynamic recrystallization. ((a) Scripta Met. [19] and (b) Metal Science [15]).

Flow Stress, Temperature, Strain Rate

Flow curves from torsion tests on high purity Al, 0.5 Mg and 0.96 Mg alloys exhibited strain hardening to plateaus which extended for strains of 0.5-2.45 [18]. The flow curves of Al deformed to strains between 3 and 60 (Figure 1) showed a peak at about 3 and a decline to a second plateau at about 20 [19-22]; an examination of the microstructure ruled out recrystallization [19,20]. For Al-4.5Mg-0.8Mn, the stress-strain (σ-ε) curves exhibited a peak at strains $\varepsilon_p \approx 1$ (Figure 1b) (which is similar to ε_p for γ - Fe and Cu, where dynamic recrystallization takes place, [6-10]) [15-17]. The microstructure after the flow softening appeared recrystallized and contained a recovered substructure characteristic of dynamic recrystallization[15-17]. Stress peaks followed by shallow declines have been observed previously in several Al alloys without evidence of dynamic recrystallization and may be due to deformational heating or to alteration in second-phase particle morphology [15,23,24]. Dynamic softening may also result from deformational heating or from texture formation at very high strains [7-15].

The dependence of peak σ_p, or steady state σ_s, stresses on temperature T and strain rate $\dot{\varepsilon}$ frequently obey the following equation [1,7,15,18]

$$A \sinh(\alpha\sigma)^n = Z = \dot{\varepsilon} \exp(+ Q_{HW}/RT) \tag{1}$$

where A, α, n and Q_{HW} are empirical constants. The results for high purity alloys are shown in Figure 2a and for commercial purity alloys in Figure 2b, with some repetition to assist comparison [13,18,23-34]. The high purity alloys (Figure 2a) exhibit a gradually increasing strength and activation energy Q_{HW} as Mg content increases with quite good agreement between differ-

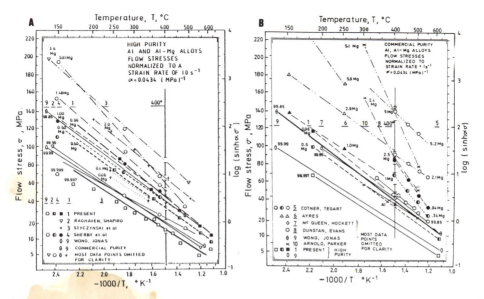

Figure 2 - Temperature dependence of the flow stress in a plot of (sinn $\alpha\sigma$) vs (-1/T)(Eqn. 1) for (a) high purity and (b) commercial purity Al and Al-Mg alloys [13,18,25-35]. The data has been normalized to $\dot{\varepsilon}$ = 1 s^{-1} and in some case data points (supporting the lines) have been omitted for clarity. ((a) Scand. J. Met. [18]).

ent authors [18,25-27,33,34]. It can also be seen that commercial Al is consistently stronger than high purity but it has about the same slope. For the commercial purity Mg alloys (Figure 2b)[28-35], there is considerable scatter in both strengths and slopes, i.e. the values of Q_{HW} (Table 1) [13,15,18,21-46]. There is not sufficient microstructural data available to explain the differences. One can only infer that they arise because of interaction between the impurities and Mg since the variations do not occur in the base Al with similar variations in impurities [18,47]. It is likely the effects are due to precipitation reactions which would explain the large variations with temperature [7,18,48-50].

The variation in hot strength with Mg content determined from Figure 2 at 400°C for both high and commercial purity alloys are plotted in Figure 3a. The plateau stress for commercial alloys at 400°C, 1s^{-1} is higher than the room temperature yield strength, yet lower than the ultimate strength which is for a lower strain; thus, while the strain hardening at 400°C is substantial, it is much less than at room temperature [18,27,47,51]. The strengths for 57S [44] 52S [52] and 5083 [15] industrial alloys at 400°C, 1s^{-1} are added and agree with the simple Al-Mg alloys quite well. For 5083 the peak stress is somewhat low, possibly the result of the dynamic recrystallization which has started taking place. For high purity alloys [25-27] comparison of the 400°C, 1s^{-1}, ε = 2.4 data with the 377°C, 10^{-3}s^{-1}, ε = 0.05 or 0.2 data [27] shows the effect of the strain hardening and of higher strain rate. From further comparison with 25°C, 10^{-3}s^{-1}, 0.005 data [27] the decrease in yield strength with rising temperature can also be appreciated. The declining slopes for all the curves indicate that the hardening effectiveness of Mg as solute declines; however, its ability to increase strain hardening is also apparent. A plot of log strength vs log at. % Mg (Fig. 3b) shows that the relation is linear with a slope of 0.5 for

199

TABLE 1   ACTIVATION ENERGIES (kJ/mol) FOR HIGH TEMPERATURE DEFORMATION

NO	TEMP. °C	$\dot{\varepsilon}$ s^{-1}	Al	0.5	1.0	1.5	2.0	2.5	3.0	3.5	4.5	5.0	6.0	IMPUR %	REFERENCES	NO
1	200-620	0.41	152	156	170									0.003	McQueen,Ryum	
2	135-520	0.12	162←(149$^+$)						209←(156$^+$)					0.05	Raghvn.Shap.	25
3	150-340	0.01	150	162		171								0.002	Styczynski,Es	26
			125$^+$	134$^+$		135$^+$									trin,Mecking	
3A	0-640	10^{-4}	120$^+$											0.002	Nicklas,Meck.	36
4	150-480	10^{-3}	126	166	171	204		265						0.001	Sherby et al	27
5	400-550	2.8	162											0.016	Cotner,	28
					256	293	139					198		0.09	Tegart	
6	150-250	10^{-3}		72	130			148					274	0.2	Ayres	29
7	200-400	0.1-220	156											0.15	McQu.Hockett	13
8	330-480	2.7-4.5	173←(152$^+$)											0.35	Dunstan,Evans	30
9	320-616	0.1-10	156											0.27	Wong,Jonas	31
9A	250-550	1.3-40	154											0.75	Alder,Phillip	32
9B	95-550	0.4-43	156											0.005	Sellars,Teg.	33
9C	204-593	10^{-7}–$^{-3}$	156											0.003	Servi,Grant	34
10	300-500	1-10	129					339				511		0.77	Arnold,Parker	35
11	300-450	10^{-4}–$^{+3}$				144$^+$								0.2	Tanaka,Nojima	37
12	300-500	0.1-1	[Al-4.5Mg-0.7Mn,5083;164]											1.3	McQueen et al	15
13	370-495	10^{-2}–$^{+1}$	[Al-4.7Mg,5356;5083;178$^+$]											0.5	Kempinnen	39
14	300-500	10^{-3}–$^{-1}$		143$^+$										0.005	Ueki,Nakamura	38
15	300-400	0.1-15	[150$^+$;Al-0.7Mg-0.85-0.5 Mn;H30]												Farag,Sellars	23
15A	400-600	4-10	150												Farag, et al.	40
16	300-540		[147$^+$;2024]												Hinesley,Con.	24
17	300-500	1.5	157$^+$	[154$^+$;M57S]										0.3	Shepp.Wright	41
18	330-490	8-50	[160$^+$;Al-0.7Mg-0.4Si]												Castle,Shepp.	42
19	300-550	2	157$^+$	[154$^+$;M57S]										0.3	Zaidi,Shepp.	43
20	300-500	0.6-100	156$^+$							[156$^+$;Al-3Mg-7Zn]				0.05	Raybld.Shepp.	44
21	300-480		163$^+$[135$^+$;2014]												Pater.Shepp.	45
22	200-500	5×10^{-2}						159				171		0.7	Painter.Pear.	47
23	400-500	10^{-3}–$^{-1}$	[155-176$^+$]						[155-190$^+$]						Perdrix et al	21

+ author's values   0.5 1.0 1.5 2.0 2.5 3.0 3.5 4.5 5.0 6.0

high purity alloys [18,25-27], 0.38 for the normalized commercial data (Figures 2,3) and 0.25 for the alloys of Cotner and Tegart [28] indicating that Mg is a less effective strengthener in less pure alloys.

The $\sinh(\alpha\sigma)^n$ function with n between 4 and 5 is used instead of the power law ($\dot{\varepsilon} \propto \sigma^n$) or the exponential law ($\dot{\varepsilon} \propto \exp\beta\sigma$) because it spans the ranges of both of these relationships giving linear plots against $\dot{\varepsilon}$ [1,15, 18,31,33,53]. These functions are drawn from creep studies where the power law with exponent between 4 and 5 has been related to dislocation climb [53-55]. The exponential law is often applied at high creep stresses where the power law breaks down, i.e. the data is no longer linear [53-55]. At the high strain rate of hot working, there is no evidence of the viscous drag behaviour with n = 3 observed in creep of Al-Mg and other solid solution alloys [53-55]. The value of α was 0.0434 MPa^{-1} for Al-Mg alloys [18] and 0.067 MPa^{-1} for 5083 alloy [15]. The value of n was 4.8 for high purity, 4.0 for low purity and 1.67 for 5083 [15,18].

Another mode of analysis which covers a broad stress range resulting from either a large variation in T or $\dot{\varepsilon}$ is the method of Kocks and Mecking [26,36,53,56,57]. In this analysis, a saturation stress σ_s^* (determined by extrapolating a plot of $\theta = d\sigma/d\varepsilon$ vs σ to $\sigma = 0$) is used with the formula

$$A(\sigma_s^*/\mu)^{n_0} = \dot{\varepsilon}\,\exp(+\,\Gamma\,\ln(\sigma_{s0}^*/\sigma_s^*)/RT) \qquad (2)$$

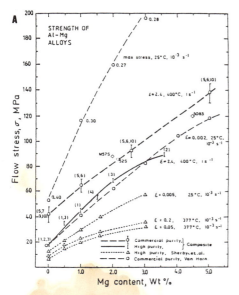

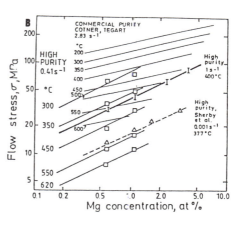

Figure 3 - The variation in strength with (a) wt % Mg and (b) at % Mg [18]. In (a) the strength at 400°C from Fig. 2 [15,28-32,35,41,52] is compared with yield and ultimate strength at 25°C [47,51] and yield and flow stresses for high purity alloys at 25 and 377°C [27]. In (b) strength depends on at % Mg raised to a power equal to the slope, based on data from Fig. 2, high purity alloys [18,25-27] and commercial alloys [28]((b) Scand. J. Met. [18]).

where n_0, $\dot{\varepsilon}_0$, σ_{s0}* and Γ are empirical constants and μ is the shear modulus. The activation energy is a function of the stress and rises to a plateau at low stress and high temperature. The values determined at high temperatures (> 350°C) were 125 and 134 kJ/mol for 0.06 and 2.0% Mg as compared to 150 and 171 kJ/mol calculated from sin $\alpha\sigma$ plots [18,26]. At lower temperatures and higher stresses, the activation energy declines more rapidly at higher Mg content and higher $\dot{\varepsilon}$ [26].

    The linearity of the strain rate dependence (Figure 3) permits determination of unique Q_{HW} values over the range 250-500°C (Table 1). The activation energy for pure Al is about 150-156 kJ/mol [1,4-7,18,26,31], but does drop as low as 126 [26,27] The commercial Al has values between 154 and 162 kJ/mol [25,28-32]. For superpurity alloys, Q_{HW} rises with Mg content to 156-166 for 0.5% Mg and to 170 for 1.0% Mg [18,26]. For higher Mg content, values of Q_{HW} rise but there is considerable difference (±20%) for different authors. For commercial alloys [28,29,32] the variation with Mg content is much greater and more irregular. The value of Q_{HW} is usually used to estimate the rate controlling mechanism. For Al the value of 150-156 kJ/mol is approximately the same as for self diffusion and is thus indicative of dislocation climb and dynamic recovery [1-7]. Slightly higher values for the alloys can be considered to result from the retarding effects of the solute on the dynamic recovery. The very high values in some cases can only be explained by precipitation at the low end of the range which greatly inhibits dislocation motion and raises the flow stress as temperature drops [48-50]. The value of 164 kJ/mol for 5083 was interpreted as indicative of dynamic recrystallization [15] but the large values for the Mg alloys tends to belie this. The traditional interpretation has been that dislocation climb was limited to a higher temperature range and below that there were

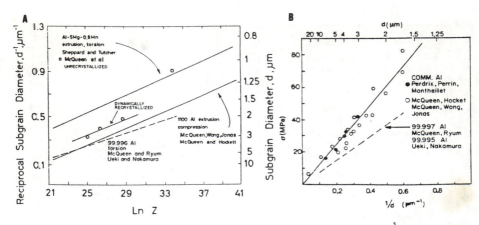

Figure 4 - The subgrain sizes produced by hot working: (a) d^{-1} vs log Z for 5083 unrecrystallized (upper) [15,58,59] and dynamically recrystallized (intermediate) [15-18], 1100 Al [13] and 99.996 Al [18,38] (b) σ_S vs d^{-1} for super and commercial purity Al [13,18,21,22,38]. ((a) Metal Science [15], (b) Mem. Et. Sci. Rev. Met. [21]).

other mechanisms such as dislocation cross-slip. In the view of Kocks and Mecking there is only one recovery mechanism for temperatures from ambient to melting and its activation energy gradually increases to that for self diffusion [26,36,56,57].

Dynamic Recovery and Substructure

The formation of subgrain structures during creep or hot-working of Al and its alloys has been much examined [1-7,11-22,25,28,42-45,54,55]. Moreover it was shown that subgrain size decreased as Z increased and σ increased (Figure 4) [1-7,13,18,21,22,25,28] and also as Mg content rose [18,25,28, 58,59]; recent work has confirmed these quantitative relationships [14-22]. In Al-4.5Mg-0.8Mn with $MnAl_6$ particles, the subgrain size was finer across the entire range of conditions [15-17,58,59]. On the other hand subgrains in 99.997 Al had the normal inverse dependence on log Z but were considerably larger than those in commercial aluminum [13,18,21,38]. It has also been shown that for T > 400°C the subgrains are equiaxed over a wide strain (0.7-3.4) although the grains are becoming increasingly elongated with strain [2-7,60]. It appears that this phenomenon is partially the result of repeated unravelling and reknitting of some subboundaries [3,4,60,61] and partially due to their migration, especially of simpler ones requiring little nonconservative motion of their individual dislocations [3,4,60,62]. The misorientation of some subboundaries is low (<1°) even at high strains (~3.4) but seems to increase to 10° for others, perhaps ones that have persisted without unravelling across large strain increments [19,21,22]. It is clear that the annihilation and rearrangement of dislocations due to dynamic recovery is responsible for the low steady state flow stresses, σ_S, independent of ε for a given Z condition. In the hot working range of temperatures and strain rates subgrains are observed in Al-Mg alloys (Figure 5) [15-18, 21,25,28,58,59]. This is consistent with the fact that the glide mechanism is not affected by viscous drag from the Mg atoms since the dislocations are moving much too fast. When viscous drag is occurring in creep (i.e. class A, alloy behaviour, n = 3), subgrains are not usually formed [54-55].

In the mechanism of disruption of subboundaries and their reformation,

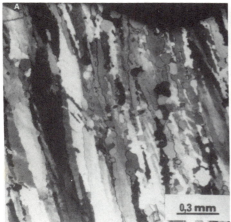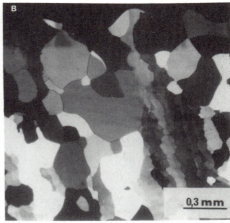

Figure 5 - Microstructure in Al-0.96 Mg deformed at 0.41 s^{-1} to a strain of 2.45 and held at temperature for a period of time (a) 450°C, 5 s, x = 5%, subgrains in unrecrystallized elongated strains (b) 550°C, 5 s, X = 80%, subgrains replaced by new grains; X 40. (Scand J. Met. [18]).

the characteristic spacing is defined by the interaction of total density, dislocation mobility and structure and stress fields of the dislocation walls [63]. This mechanism is not particularly noticeable in simple, constant stress creep tests because the steady state strain is rather low and hence the limited elongation of the grains is insufficient to contrast with the dynamic equiaxiality of the subgrains [2-5,57,60]. Although the unravelling and reknitting of the subboundaries in Al-Mg alloys at elevated temperatures has been observed directly [61], the lack of elongation of the subgrains is frequently ascribed to dislocation wall migration, although observations indicate that it contributes only 12% of the strain [62]. The creep phenomenon currently explained by the disruption-reformation mechanism is the change in subgrain size which accompanies a sudden change in stress or temperature [54,64-68]. If the stress is lowered or the temperature raised, the number of walls gradually decreases as there is a net disintegration and a formation of new walls at the equilibrium spacing. In the opposite change of conditions, there is a net formation of walls to give a spacing less than the pre-existing one.

This process of unravelling and reknitting can be associated with the Kuhlman-Wilsdorf model for strain hardening in which the flow stress is defined by the force needed to bow segments out of the cell walls [69-70]. This force increases as the link length decreases with build up in the dislocation density. Formerly, bowing-out from the walls was dismissed as a source during creep in favor of that from the network within the cells, where the link length is much longer and the force correspondingly less. However, recent internal stress measurements by transient creep tests indicate that within the subgrains there is a backstress of 0.5 σ_s and within the thickness of the walls a forward stress of about 5 σ_s. This differential of 10 σ_s greatly enhances the operation of wall sources over those in the internal network [64-68]. The existence of these internal stresses is also supported by observations of static recovery in which they greatly enhance the initial rate at which the redundant dislocations, in and near the walls, annihilate in comparison to a later time after the stresses are relaxed [71,72]. Application of a small forward or backward stress markedly speeds up or slows down, respectively, the early progress of this first stage

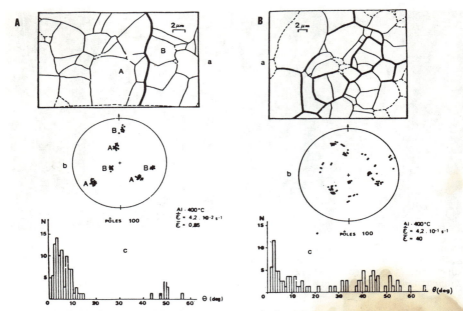

Figure 6 - Orientation differences across subgrain walls were determined by diffraction (a) for ε = 0.85 and (b) for ε = 40 [21,22]. In (a) there is one grain boundary separating large regions of subgrains with misorientations averaging about 5°C, but in (b) for about 4 cellularized grains, there are many high angle boundaries between 10 and 40° arranged in an irregular way (Mem. Et. Sci. Rev. Met. [21].

due to reinforcing or countering the internal stresses in the walls [72].

As a result of the above findings, the following qualitative model for steady state hot deformation at stress σ_s is proposed. Dislocations are generated by the bowing out of segments from the sub-boundaries and from the network within the subgrains [3,4,61,67,69,70]. These expanding segments, encountering only minor obstacles within the subgrains, glide, cross-glide and climb until they lodge near a sub-boundary in a ragged array, a sort of pseudo pile-up, until they are able to knit into it [61]. These ragged arrays, and the "flexure" of the walls by the applied stress exert a back stress of about 0.5 σ_s within the subgrain [64-68]. The ratio of back to applied stress has been shown to remain constant as stress increases, which is consistent with the rising raggedness and density of the walls [64-67]. The dislocations at the walls also create a forward stress concentration within the subboundaries of about 5 σ_s [64-67], which is able to bow out wall links much shorter than those of the network. Recovery, i.e. an increase in link length, takes place in the walls by two mechanisms. Annihilation occurs when oppositely signed dislocations, which have knit themselves into the subboundary from opposite sides, climb and glide until they coincide [61]. Unravelling of the subboundary as dislocations are pushed out of it also increases the link length [61,69-72]. In steady state, these mechanisms balance the hardening from new dislocations knitting themselves into the subboundaries.

When Al is torsionally deformed to strains of 3 to 60, the grains are twisted into thin helicoids which spiral around the specimen; the grain thickness in the axial direction decreases inversely as the equivalent

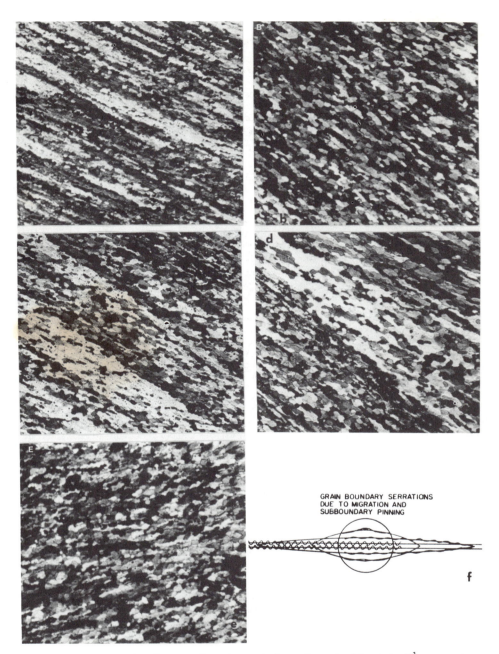

GRAIN BOUNDARY SERRATIONS
DUE TO MIGRATION AND
SUBBOUNDARY PINNING

f

Figure 7 - 99.9% Al of 2 grain sizes deformed at 400°C, 0.2 s^{-1} to various strains and examined in polarized light [19,20] (a) 0.1 mm, ε = 3, x 50; (b) 0.1 mm, ε = 40, x 75; (c) 2 mm, ε = 10, x 50; (d) 2 mm, ε = 60, x 75, reduced subgrain contrast; (e) as (d) rotated to high subgrain contrast x 75; (f) schematic of change in grain shape and increase in serration amplitude. (Scripta Met [19]; Practical Metallography [20]).

strain [19-20]. At 400°C, 0.41 s^{-1} for strains between 4 and 40 [21,22], the subgrain size increased by about 10% in relationship to the decline in flow stress with the formation of a preferred orientation. The additional alteration in substructure was the presence of many high angle boundaries along with some remaining low angle boundaries between the subgrains or crystallites (Figure 6) [21,22]. Moreover, the distribution of the high angle ones was quite irregular compared to the simple organization of grain boundaries surrounding large regions divided by subboundaries at ε = 0.7-4 [3-5,7,13,60]. When experiments were carried out at 400°C, 0.2 s^{-1} with 0.1 and 2 mm grain diameters, the large grains could be traced in elongated form up to ε = 60 on tangential sections [19-20]. Polarized light, when away from extinction, gives each grain a uniform color or shade (Figure 7). The boundaries, which can also be distinguished by topographical contrast in ordinary light, are quite serrated as a result of local migrations in association with subgrain formation [4,5,19,20,73]. The subgrains are evident at all strain being in sharp contrast when the polarized light comes close to extinction (Figure 7). On transverse sections, the grains remained equiaxed while subgrains developed; as the strain increased to 40 and 60, subgrains of high misorientation appear within the original grains apparently due to penetration of serrations on grains from above and below as the grains become very thin [j]. This effect is also observed in tangential sections where the grains are pinched off by serrations from grains on opposite sides meeting.

In the specimens with fine grains (the same as those whose substructures at ε = 4 and 40 [21,22] were described above), the grains can be traced easily to strains of 10 where they are becoming as thin as the subgrain diameter [19-20]. At 20 and 40, only short segments of grains, i.e. rows of a few similarly oriented subgrains, can be distinguished presumably as a result of pinching off by bulges from grains on opposite sides. The transverse sections give evidence of the original grains but with numerous interpenetrations as the grain thickness becomes thinner than the subgrain diameter. At ε = 60, the subgrains of neighboring grains have so interpenetrated that they appear almost randomly oriented and no vestiges of the grains are discernible. The orientations of the extinction directions of grain in different sections show that the grains are rotating into a common alignment as the strain increases; such texture was confirmed by X-ray diffraction. Subgrain size, as determined by polarized light, remains approximately constant with strain (Figure 8) [19]. The substructures were also examined by channeling contrast SEM which gives similar appearance to the polarized light (Figure 9)[19-20]. The subgrains and grains are distinguishable to ε = 60 in the coarse grained specimen, but by ε = 40 the grains are difficult to find in the 0.1 mm one. TEM examination of the substructure showed that the subgrains remain constant in size with strain and are the same for both grain sizes consistent with the similarity in flow stress (Figure 1a). At all strains, low angle boundaries containing simple arrays of dislocations are observed (Figure 9) [19-20]. The subgrains remain approximately equiaxed throughout these large strains although the grains become extremely elongated, with only limited migration of the high angle boundaries. At strains of 40 or 60, the grains essentially become very long ribbons with thickness near or below that of the subgrains which are formed by dislocation walls mostly normal to the serrated grain boundaries. The material continues to deform at constant stress even though the subgrains change from having mainly low misorientation boundaries to having almost half high-angle ones. This crystallite structure, defined by boundaries almost equally mixed high-and low-angle, and having the same characteristic size as subgrains at low stresses [19-22], has not formed by either discontinuous, or continuous, dynamic recrystallization. One could hypothesize that the mechanism is a combination of dislocation motion with dynamic recovery and of grain boundary sliding.

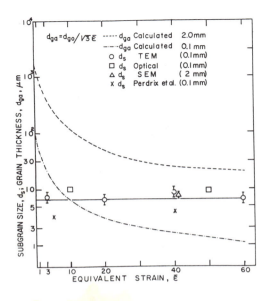

Figure 8 - Calculated axial grain thickness for 2 and 0.1 mm grains as a function of equivalent strain ($\gamma = \sqrt{3}\,\varepsilon$). Subgrain sizes measured by polarized light, SEM and TEM for deformation at 400°C, 0.2 s^{-1} [19] with some comparative data [21]. (Scripta Met. [19]).

Figure 9 - (a) SEM micrograph of Al deformed at 400°C, 0.2 s^{-1} to $\varepsilon = 60$, 2 mm grains containing subgrains still distinguishable.
(b) TEM micrograph for $\varepsilon = 60$ showing the dislocation array in a low angle boundary [19,20].

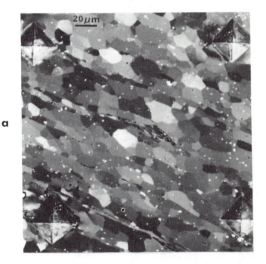

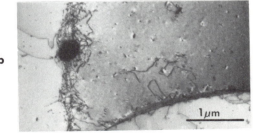

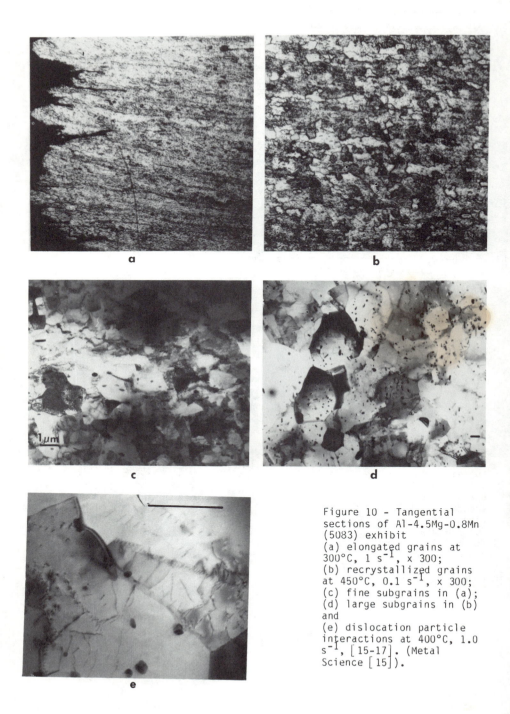

Figure 10 - Tangential sections of Al-4.5Mg-0.8Mn (5083) exhibit
(a) elongated grains at 300°C, 1 s⁻¹, x 300;
(b) recrystallized grains at 450°C, 0.1 s⁻¹, x 300;
(c) fine subgrains in (a);
(d) large subgrains in (b) and
(e) dislocation particle interactions at 400°C, 1.0 s⁻¹, [15-17]. (Metal Science [15]).

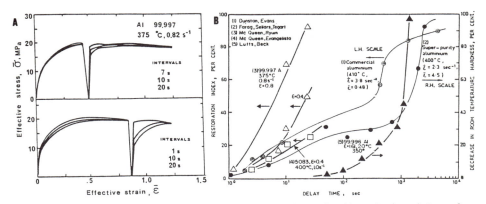

Figure 11 - (a) Static restoration determined by reloading during intervals after strains of either 0.4 or 0.8 at 375°C, 0.8 s^{-1} [18]; (b) The fractional softening or restoration index from hot working and cold-working-anneal- [1,15,18,30,40,78] ((a) Scand. J. Met. [18], Int. Met. Rev. [1]).

In contrast with the behavior of Al where the original grains persist to extreme strains, Al-4.5Mg-0.8Mn exhibits equiaxed grains (Figure 10) after strains of 5 when T exceeds 400°C and $\dot{\varepsilon}$ is less than 1 s^{-1} [15-17]. The new grains contain a subgrain structure (Figure 10) which appear qualitatively consistent in dimension and dislocation arrangement with the substructures at higher Z [7-10,15,58,59]. However, more precise examination shows that the subgrains in the dynamically recrystallized grains are larger than those in non-recrystallized specimens (Figure 4a). The 5083 alloy contains MnAl6 in the form of constituent particles (>1 μm) and eutectic rods (<0.2 μm) fractured by the deformation. The eutectic particles pin dislocations and frequently are situated in subboundaries, their density being sufficient to retard recrystallization [16]. Dynamic recrystallization appears to result from enhancement of nucleation by the large constituent particles [15-17,58,59]. (The influence of particles with different sizes and distributions [16,74-76] is discussed in a paper on dilute Al-Fe conductor alloys in this book [76]). As in static recrystallization, the nuclei grow in the regions of fine highly misoriented cells which have formed around these large undeforming particles [74,75]. These grains initially grow rapidly but are stopped at a fairly small size by the fine particles. Of course the dynamically recrystallized grains are themselves deformed, rapidly developing the characteristic substructure [2-10,15-17] and later recrystallizing again.

Static Recovery and Recrystallization

The hot working process includes not only the pre-heating and the one or several stages of deformation, but also the periods between stages and during cooling [2,8,77]. Static restoration takes place during any hold at elevated temperature and progresses from recovery to recrystallization as time increases. In the final cooling it may be desirable to retain the hot-work substructure as a means of strengthening in subsequent drawing or in service [76]; but in other cases a recrystallized structure improves the ductility or the stress corrosion resistance [11,12]. Softening between passes reduces the flow stress in the next pass and usually increases the hot ductility. The total restoration can be measured by mechanical testing at room temperature and the recrystallization by metallography. The former can also be determined by mechanical metallography at the deformation temp-

209

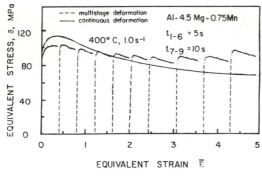

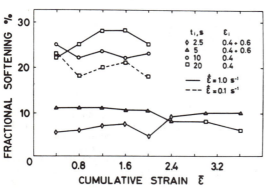

Figure 12 - (a) Multi-stage tests on Al-4.5Mg-0.8Mn, at 400°C, 1 s^{-1} with ε_n = 0.4 and 0.6 with t_n = 10 s; (b) Fractional softening plotted against cumulative ε is greater for longer times, higher ε_n and higher $\dot{\varepsilon}$ [15] (Metal Science).

erature (Figure 11)[8,15,18,30,40,48-50,77,78], by reloading after a holding period and calculating the fractional softening FS after the nth pass as

$$FS_n = (\sigma_{un} - \sigma_{r(n+1)})/(\sigma_{un} - \sigma_{Y1}) \qquad (3)$$

where σ_u is the stress at unloading before the hold, σ_r on reloading and σ_Y at yielding in the first pass. Durrstan and Evans [30] used this technique to show that in Al at about 400°C 30-40% softening took place by recovery within 100 s but complete recrystallization (FS, 100%) required about 1000.

The softening in holds of constant time between a series of equal passes (0.4) was determined mechanically for Al-4.5Mg-0.8Mn (Figure 12) [15]. The average softening at 400°C (0.72 Tm), 1.0 s^{-1} increased from about 5% at 2.5 sec to about 28% at 20 s; the FS was lowered to 18% when the strain rate was reduced to 0.1 s^{-1}. Under similar circumstances, the softening in pure aluminum is 50% in 20 s [18] which indicates the stabilizing influence of the fine MnAl$_6$ particles in 5083. It is interesting to note that at 950°C (0.68 Tm), 0.1 s^{-1} with ε_n = 0.2 the softening was about 100% for C steel but only 25% in HSLA steel with Nb(C,N) precipitates [48], 40% in stainless steel with 29% solute [49], or 10% in M2 tool steel with 17% carbide [50]. It is evident that even with the static recovery the 5083 strain hardens back up to the characteristic plateau by the end of the pass. The grain and substructures in specimens quenched at the end of deformation are the same for both continuous and interrupted testing [15]. Thus, in the normal holds between passes in hot rolling of Al alloys only limited recovery takes place and the reduction in the mean pass flow stress is only a few percent. In interrupted testing of Al-Mg alloys, Cotner and Tegart determined that the ductility in interrupted tests was almost doubled in comparison to that in continuous tests [8,28]. The cause of this appears to be the static recovery during the holds which additionally softens the grains helping them to accommodate to stress concentrations at triple junctions due to grain boundary sliding. However, an interruption sequence which permits recrystallization between passes reduces the ductility presumably because it prevents the formation on the grain boundaries of heavy serrations which normally reduce the grain boundary sliding and hence the triple point cracking [8,28]. Also of significance in multiple stages, the flow curve in a pass is progressively lower as more recovery or recrystallization takes place prior to the pass and, in the limit of 100% recrystallization, reduces to identity with the initial pass curve [8,30,77]. In a multistage test with declining temperature (as occurs industrially), where there is complete recrystallization between passes, the flow curves would be successively higher just as if each pass had been independently determined. However, if recrystallization does not take place between two passes then the flow curve at the lower temperature would be lower than in the recrystallized case because the dislocations in the inherited high temperature substructure would be able to move about more easily during a transient stage in which the substructure is altering to that characteristic of the lower temperature condition [1,2,8,40]

At room temperature, the strength of material retaining the hot-worked substructure is usually related to the subgrain diameter by a modified Hall-Petch relationship:

Figure 13 - Subgrain structure in 1100 Al after compression to $\varepsilon=0.7$:at 400°C 0.1 s^{-1},(a) as worked TEM (b) after 200 hrs at 200°C SEM; and at 200°C, 1.4 s^{-1}, (c) as worked TEM (d) after annealing SEM [14] (Riso National Lab.).

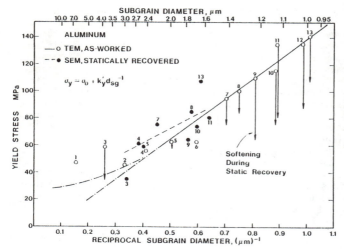

Figure 14 - Dependence of room temperature strength on subgrain size in 1100 Al either after hot-working to ε = 0.7 at various T and ε̇ or after annealing at 200°C for 200 hrs following hot-working [14] (Riso National Laboratory).

$$\sigma_Y = \sigma_0 + k_Y d^{-p} = \sigma_0 + k_Y(d^{\frac{1}{2}-p})d^{-\frac{1}{2}} \qquad (4)$$

where p is usually unity [2-5,8,11,14,76,79]. The subboundary strengthening coefficient $k_Y(d^{\frac{1}{2}-p})$ thus increases as d becomes smaller, consistent with the increase in wall density observed in TEM. There is some evidence that statically-recovered cold-worked substructures exhibit $p = \frac{1}{2}$ [3,80]. To test this hypothesis, the hot-worked material was statically recovered for 200 hrs at 200°C (avoiding recrystallization) to study the effect on substructure and strength [14]. Examination of bulk specimens by SEM channeling contrast revealed a substructure qualitatively similar to those observed by TEM in thin foils (Figure 13) [14] (this confirms that TEM subgrains are not simply thinning artifacts). The static annealing resulted in a decrease in strength (Figure 14) and an increase in subgrain size; the changes were greatest for those worked at the lowest temperature and highest strain rate. The SEM microscopy did not permit analysis of any changes in dislocation density or configuration in the subboundaries. The final relationship between strength and diameter remained almost the same with p = 1 but k_Y slightly reduced. It should be noted that extrapolation to higher diameters cannot be linear (this would give a negative σ_0) but must curve towards horizontal as subboundaries become very diffuse [7,79].

The progress of recrystallization after hot-working to a strain of 2.45 was determined for Al, 0.5 Mg and 1.0 Mg alloys over the range 350 to 620°C by simply holding for various times before quenching. The undistorted, equiaxed, recrystallized grains were easily distinguished from the elongated grains containing well developed hot-work subgrains (Figure 5) [18]. The time-temperature-fraction recrystallized curves in Figure 15a were determined from the traditional s-curves at each temperature. The evident retarding effect of the Mg solute appears to be about the same for both concentrations. This effect of Mg is consistent with the cold-work-annealing results of Perryman [81] who found that those concentrations gave almost equal large retardations whereas larger ones gave increasingly less retardtion (because of the greatly increased strain energy induced by the higher level of Mg content [18,82]). At 350°C the time for recrystallization after cold-working was higher by a factor of 10^3 although the stored strain energy was much higher as may be inferred from the strength curves in Figure 3 [18]. This effect may arise from the loss in strain energy during heat up to the annealing temperature which is absent in the hot-work tests where the

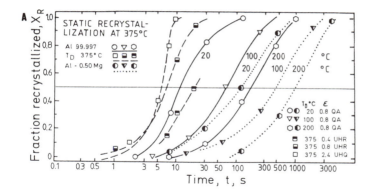

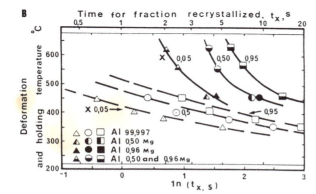

Figure 15 - (a) Fraction recrystallized at 375°C from Unloading Holding and Quenching (UHQ, microscopy), or Reloading (UHR), tests are advanced compared to Quenching and Annealing tests. The 0.5 Mg alloy is retarded compared to pure Al. (b) Time-temperature-recrystallization curves for Al and for 0.50 and 0.96 Mg. (Scand J. Met [18]).

deformation ends at the annealing temperature. This effect is also seen in the retardation of the quench-anneal curves compared to unload-hold curves at 375°C in Figure 15a [18,82]. For lower purity alloys, the recrystallization time at 350°C after hot-working to a lower strain of 0.73 at 2.45 s^{-1} was a factor of 10^2 greater, but was somewhat less for the 1% than for the 0.5% Mg alloy [28]. In those alloys, the time for recrystallization decreased with Mg content up to 5% in agreement with Perryman's results [28,81]. The variation in recrystallization rate with temperature is much less than after a fixed cold strain because the strain energy induced decreases with increase in the common working-holding temperature. This effect can be seen in the slower recrystallization as the deformation temperature is raised from 20 to 200°C (Figure 15a) [18]. Mg alloys have a lower dependence on temperature because the diminished boundary mobility is important at high temperatures but the increased strain hardening is significant at low. The effect of increasing strain energy at constant T is seen in specimens worked to different ε and annealed at 375°C (Figure 15a) [18].

The recrystallization fraction X from the S curves was plotted according to the Avrami relation

$$X = 1 - \exp(-Bt^k) \qquad (5)$$

where B and k are constants and t is time (Figure 16)[18]. In the above hot working and holding experiments with microscopic determination, the value of k was 3.0 for Al and 3.6 for the 0.5 and 1.0 Mg alloys [18]. These values are higher than Perryman's 1.14, 1.60 and 1.43 although he reports previous

213

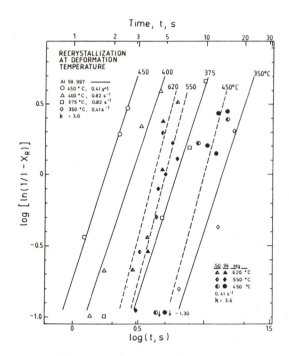

Figure 16 - Avrami plot for recrystallization of 99.997 Al and alloys of 0.5 and 0.96 Mg held at the working temperature before quenching and measuring by microscopy (Scand J. Met [18]).

values between 3.3 and 5.5 [81]. For zone refined Al, which recovered appreciably as recrystallization progressed, k was equal to 2.2 [83]. Similar experiments on Cu yielded k = 2.2 [84]; however, the strain was much lower and nucleation appeared to be more confined to grain boundaries. For specimens worked at 20-200°C and then annealed at 375°C, thus similar to Perryman's cold work and anneal, the value of k is 1.1 for Al and alloys [18]. From the results of mechanically measured restoration which includes recovery, k = 1.6 compares well with 1.3 for Cu similarly studied [18,84].

Conclusions

The dynamic restoration mechanism in Al up to extreme strains is recovery, although the frequency of high angle boundaries on subgrains rises to about 50% as the grains thin down to the diameter of the subgrains. Mg up to 5% retards the recovery and reduces the subgrain size to a small degree which is not sufficient at 1% Mg to accelerate recrystallization by countering the retarding effect on boundary mobility. In alloys with dispersoids, dynamic recrystallization occurs from enhanced nucleation at particles larger than 0.6 μm whereas small ones (~0.2 μm) pin dislocations and stabilize the subboundaries. The sinh strain rate and Arrhenius temperature dependencies are satisfactory. The activation energy is near that for self diffusion for Al and dilute Mg alloys. However, it rises as particles cause dynamic recrystallization and Mg related precipitates occur at high Mg concentrations. Static recovery and recrystallization follow hot-working if the temperature is maintained and have essentially the same behaviour as in annealing after cold deformation.

Acknowledgements

We wish to acknowledge the collaboration in research which provides the foundation of this review, specifically work with W.B. Hutchinson, E.

E. Evangelista, H. Mecking, N. Ryum, E. Nes and J. Solberg. Moreover we are grateful for the funding which made the collaboration possible i.e. grants from the national research councils of Canada, Italy, West Germany and Norway. Finally we thank J. Bowles for assistance in metallography and preparation of diagrams.

References

1. C.M. Sellars and W.J. McG.Tegart - Hot workability, Int. Met. Rev. 17, (1972), 1-24.

2. H.J. McQueen and J.J. Jonas - Recovery and recrystallization during high temperature deformation, Plastic Deformation of Materials (Treatise on Mat. Sci. Tech., Vol. 6), (Ed. R.J. Arsenault), pp. 393-493, Academic Press, New York (1975).

3. H.J. McQueen - Production and utility of recovered dislocation substructures, Met. Trans. 8A (1977), 807-24.

4. H.J. McQueen - Dynamic recovery and its relation to other restoration mechanisms, Met. i Odlewnictwo 6, (1979), 421-450.

5. H.J. McQueen - Discovery of the dynamic restoration mechanisms in hot working, Metal Forum (Aust) 4 (1981), 81-91.

6. H. Mecking and G. Gottstein: Recovery and recrystallization during deformation, Recrystallization of Metallic Materials (ed. F. Haessner), pp. 195-212, Dr. Reiderer Verlag Stuttgart, (1977).

7. H.J. McQueen and J.J. Jonas - Recent advances in hot working - fundamental dynamic softening mechansims, J. Appl. Metal Working 4 (1984), 233-241.

8. H.J. McQueen and J.J. Jonas - Dynamic and static softening mechanisms in multistage hot working, J. Appl. Metal Working 3 (1985), 410-420.

9. C.M. Sellars - Recrystallization in metals during hot deformation, Phil. Trans. R. Soc. Lond, A288 (1978), 147-158.

10. T. Sakai and J.J. Jonas - Dynamic recrystallization: mechanical and microstructural considerations, Acta Met. 32 (1984), 189-209.

11. H.J. McQueen - Experimental roots of thermomechanical treatments for Al-alloys, J. Met. 32 [2] (1980), 17-36.

12. E.A. Starke and and J.C. Williams - Role of thermomechanical processing in tailoring properties of Al and Ti alloys, Deformation, Processing and Structure, pp. 279-354, ASM, Metals Park, Ohio (1984).

13. H.J. McQueen and J.E. Hockett - Microstructures of Al compressed at various rates and temperatures, Met. Trans. 1 (1970), 2997-3004.

14. H.J. McQueen and W.B. Hutchinson, SEM of recovered substructures in Al, Deformation of Polycrystals (ed. N. Hansen et al.), pp. 335-342, Riso Natl. Lab., Roskilde, Denmark (1981).

15. H.J. McQueen, E. Evangelista, J. Bowles and G. Crawford - Dynamic

recrystallization in Al-5Mg-0.8Mn, MET. Sci. 18 (1984), 395-402.

16. E. Evangelista, H.J. McQueen and E. Bonetti - Interaction between (MnFe)Al$_6$ particles and substructure formed during hot working of Al-5Mg-0.8Mn alloy, Deformation of Multi-Phase and Particle Containing Materials (ed. J. Bilde-Sorensen et al.), pp. 243-250, Riso Natl. Lab., Roskilde, Denmark (1983).

17. E. Evangelista, E. Bonetti, R. Tognato and H.J. McQueen - Recovery and recrystallization in a hot-worked comm. Al-Mg-Mn alloy. Gaz. Chim Ital. 113 (1983), 305-307.

18. H.J. McQueen and N. Ryum - Hot working and subsequent static recrystallization of Al and Al-Mg alloys, Scand. J. Met. 14 (1985), (in press).

19. H.J. McQueen, O. Knustad, N. Ryum and J.K. Solberg - Microstructural evolution in Al deformed to strains of 60 at 400°C, Scripta Met. 19 (1985), 73-78.

20. O. Knustad, H.J. McQueen, N. Ryum and J.K. Solberg - Polarized light observation of grain extension and subgrain formation in Al deformed at 400°C to very high strains, Practical Metal. 22 (1985),(in press, May).

21. Ch. Perdrix, M.Y. Perrin and F. Montheillet - Comportement mécanique et évolution structural de l'Al au cours d'une deformation à chaud de grande amplitude, Mem. Et. Sci. Rev. Métal 78 (1981), 309-320.

22. F. Montheillet - Comportement mécanique et structurale des materiaux à forte énergie de défaut d'empilement sous grande deformation à chaud, Les Traitments Thermoméchanique (ed, P. Costa et al.), pp. 57-70, INSTN, Saclay (1981).

23. M.M. Farag and C.M. Sellars - Double max. flow patterns in extrusion of H30 Al-alloy, Met. Tech. 2 (1975), pp. 220-228.

24. C.P. Hinesley and H. Conrad - Effect of T and $\dot{\varepsilon}$ on flow in extrusion of 2024 alloy, Mat. Sci. Eng. 12 (1973), pp. 47-58.

25. M. Raghaven and E. Shapiro - Structure and strengthening of Al-Mg alloys during hot working, Met. Trans. 11A (1980), 117-121.

26. A. Styczynski, Y. Estrin and H. Mecking - High temperature deformation of Al-Mg alloys. Creep and Fracture of Engineering Materials and Structures (ed. B. Wilshire and D. Owen), pp. 115-130, Pineridge Press, Swansea, U.K. (1984).

27. O.D. Sherby, R.A. Anderson and J.E. Dorn - High temperature strength of high purity Al-Mg alloys, Trans. AIME 191 (1951), 643-652.

28. J.R. Cotner and W.J. McG. Tegart - High temperature deformation of Al-Mg alloys, J. Inst. Metals 97 (1969), 73-79.

29. R.A. Ayres - Alloying Al with Mg for ductility at warm temperatures (25-250°C), Met. Trans. 10A (1979), 849-854.

30. R.W. Evans and G.R. Dunstan - Hot working and subsequent restoration of commercial purity Al, J. Inst. Metals 99 (1971), 4-14.

31. W.A. Wong and J.J. Jonas - Al extrusion as a thermally activated process, Trans. Met. Soc. AIME 242 (1968), 2271-2280.

32. J.F. Alder and V.A. Phillips - Effect of temperature and stain rate on resistance of Al, Cu and Fe to compression, J. Inst. Metals 83 (1954-55), 80-86.

33. C.M. Sellars and W.J. McG. Tegart - The relation between strength and structure in hot deformation, Mém. Sci. Rev. Mét. 63 (1966), 731-746.

34. I.S. Servi and N.J. Grant - Structural observations on Al deformed in creep, Trans. AIME 191 (1951), 909-916.

35. R.R. Arnold and R.J. Parker - Resistance to deformation of Al and some Al alloy, J. Inst. Metals 88 (1960), 255-259.

36. B. Nicklas and H. Mecking - Analysis of steady-state flow in Al single crystals, Strength of Metals and Alloys (ICSMA 5) (ed. P. Haasen et al), Vol. 1, pp. 351-356, Pergamon Press, Oxford (1983).

37. K. Tanaka and T. Nojima - Effects of T and $\dot{\varepsilon}$ on strength of Al-1.1Mg, Proc. 20th Japan Congress Mat. Res. pp. 89-93 (1977).

38. M. Ueki and T. Nakamura - Substructural strengthening by hot working in two FCC materials, Trans. JIM 17 (1976), pp. 139-148.

39. A.J. Kemppinen - Simulation of hot working characteristics of Al-Mg alloys by high strain rate tensile testing, Deformation Under Hot Working Conditions, pp. 117-121, Iron Steel Inst., London (1968).

40. M.M. Farag, C.M. Sellars and W.J. McG. Tegart - Simulation of hot-working of Al, Deformation Under Hot Working Conditions (SR 108), pp. 60-67,101, Iron Steel Inst., London (1968).

41. T. Sheppard and D.S. Wright - Determination of constitutive equation for Al alloys at elevated T., Met. Tech. (1979), 215-223.

42. A.F. Castle and T. Sheppard - Development of product structure in extrusion of Al alloys Met. Tech. 3 (1976) 433-436, 454-464, 465-475.

43. M.A. Zaidi and T. Sheppard - Development of microstructure through roll gap in rolling of Al alloys, Met. Sci. 16 (1982), 229-238.

44. D. Raybould and T. Sheppard - Thermally activated plastic flow in mechanical working processes, J. Inst. Met. 101 (1973), 45-52.

45. T. Sheppard and S.J. Paterson - Direct and indirect extrusion of Al, Met. Tech. 9 (1982), 274-281.

46. M.J. Painter and R. Pearce - Hot forming of Al-Mg alloy sheet, Formability of Metallic Materials - 2000 AD (ed. J.R. Newby), pp. 105-118, ASTM, Philadelphia (1982).

47. L.F. Mondolfo - Aluminum Alloys: Structure and Properties, pp. 806-841, Butterworths, London (1976).

48. J. Sankar, D. Hawkins and H.J. McQueen - Torsion simulation of rolling of C and HSLA steels, Met. Tech. 6 (1979), 325-331.

49. N.D. Ryan, H.J. McQueen and J.J. Jonas - Deformation behavior of 304, 316 and 317 stainless steels during hot torsion, Can. Met. Q. 22 (1983), 369-378.

50. C. Imbert, N.D. Ryan and J.J. Jonas - Hot workability of tool steels, Met. Trans. 15A (1984), 1855-1864.

51. K.R. van Horn - Aluminum, Vol. 1, Properties, Physical Metallurgy and Phase Diagrams, P. 196, ASM, Metals Park, Ohio (1967).

52. R. Horiuchi, J. Kaneko, A.B. Elsebai and M.M. Sultan - Characteristics of Hot Torsion for Assessing Hot Workability of Al Alloys, Inst. Space Aero. Sci. Report, (Univ. Tokyo No. 443), 35 [1] (1970) 1-19.

53. H.J. McQueen and H. Mecking - Comparison of deformation and failure mechanism observed in hot working and creep of metals, Creep and Fracture of Engineering Materials and Structures, pp. 169-84, Pineridge, Press, Swansea, U.K. (1984).

54. W.D. Nix and B. Ilschner - Mechanisms controlling creep of single phase metals and alloys, Strength of Metals and Alloys (ICSMA 5) (ed. P. Haasen et al), Vol. 3, pp. 1503-1530, Pergamon Press, Oxford (1979).

55. T.G. Langdon - Deformation at high temperatures, Strength of Metals and Alloys (Ed. R.C. Gifkins),pp. 1105-1120, Pergamon Press, Oxford (1982).

56. U.F. Kocks - Laws for work-hardening and low temperature creep, J. Eng. Mech. Tech. (ASME H) 198 (1976), 76-85.

57. H. Mecking - Strain hardening and dynamic recovery, Dislocation Modelling of Physical Systems (ed. M.F. Ashby et al.), pp. 197-211, Pergamon Press, Oxford (1981).

58. T. Sheppard, M.G. Tutcher and H.M. Flower - Recovered dislocation substructure in Al-5Mg, Met. Sci. 13 (1979), 473-481.

59. T. Sheppard and M.G. Tutcher - Effect of processing on structure and properties of Al-5Mg-0.8Mn, Met. Tech. 8 (1981), 319-327.

60. H.J. McQueen, W.A. Wong and J.J. Jonas - Deformation of Al at various $\dot{\varepsilon}$ and T., Can. J. Phys. 45 (1967), 1225-1234.

61. V.K. Lindroos and H.M. Miekk-oja - Structure and formation of subboundaries in Al-Mg alloys, Phil. Mag. 16 (1967), 593-610; 17 (1968), 119-33.

62. S.F. Exell and D.H. Warrington - Subgrain boundary migration in Al, Phil. Mag. Ser. 8, 26 (1972), 1121-1136.

63. D.L. Holt - Dislocation cell formation in Metals, J. Appl. Phys. 41 (1970), 3197-3201.

64. W. Blum - On modelling steady state and transient deformation at elevated temperataure, Scripta Met. 16 (1982), 1353-1357.

65. W. Blum and H. Schmidt - Steady state and transient creep of Al at 400 K, Res Mech. 9 (1983), 105-124.

66. W. Blum and A. Finkel - New technique for evaluating long range internal black stresses, Acta Met. 30 (1982), 1705-1715.

67. W. Blum, H. Munch and P.D. Portella - Application of knitting model of creep to pure Al Al-Mg Al-Zn alloys, Creep and Fracture of Engineering Materials and Structures (ed. B. Wilshire, D. Owen), pp. 131-147,

Pineridge Press, Swansea, U.K. (1984).

68. W.D. Nix, J.C. Gibeling and K.P. Fuchs - Role of long range internal back stresses in creep of metals, Mechanical Testing for Deformation Model Development, STP 765 (ed. R.W. Rhode et al.) p. 301 - ASTM Philadelphia (1982).

69. D. Kuhlman-Wilsdorf - Unified theory of stage II - III work hardening in pure fcc metals, Work Hardening (ed. J.P. Hirth), pp. 97-132, Gordon and Breach, N.Y. (1968).

70. D. Kuhlman-Wilsdorf - Recent progress in understanding of pure metal and alloy hardening, Work Hardening in Tension and Fatigue (ed. A.W. Thompson), pp. 1-44, AIME, Warrendale, PA. (1977).

71. J. Hausselt and W. Blum - Dynamic recovery during and after steady state deformation of Al-11%Zn, Acta Met. 24 (1976), 1027-1039.

72. T. Hasegawa, T. Yakou and U.F. Kocks - Length changes and stress effects during recovery of deformed Al, Acta Met. 30 (1982), 235-243.

73. F.J. Humphreys - Inhomogeneous deformation of some Al alloys at elevated temperature, Strength of Metals and Alloys (ICSMA 6) (ed, R.C. Gifkins), Vol. 3, pp. 625-630, Pergamon Press, Oxford (1982).

74. F.J. Humphreys, Nucleation of recrystallization at second phase particles in deformed Al, Acta Met. 25 (1977), 1323-1344.

75. E. Nes - Recrystallization with bimodal particle distributions, Recrystallization and Grain Growth, of Multiphase, Particle Containing Material, pp. 85-95, Riso Natl. Lab., Roskilde, Denmark (1980).

76. H.J. McQueen, H. Chia and E.A. Starke, Microstructural strengthening mechanisms in dilute Al-Fe conductor alloys, Microstructural control in Al Alloy Processing, (ed. H. Chia and H.J.McQueen) (in press) AIME Warrendale, PA (1985).

77. H.J. McQueen, M.G. Akben and J.J. Jonas - Mechanical metallography in hot working, Microstructural Characteristics of Materials by Non-Microscopic Techniques, pp. 397-404, Riso Natl. Lab., Roskilde, Denmark (1985).

78. A.H. Lutts and P.A. Beck - Annealing of a cold rolled Al single crystal, Trans AIME 200 (1954), 257-260.

79. D.J. Abson and J.J. Jonas - The Hall-Petch relationship and high temperature subgrains, Met. Sci 4 (1970), 24-28.

80. R.J. McElroy and Z.C. Szkopiak - Dislocation-substructure strengthening and mech.-ther. treatment of metals, Int. Met. Rev. 17 (1972), 175-202.

81. E.C.W. Perryman - Recrystallization characteristics of superpurity base Al-Mg alloys, Trans. AIME 203 (1955), 369-378.

82. N. Ryum and J.D. Embury - Recrystallization in aluminum, Recovery and Recrystallization of Metals (Ed. L. Himmel), pp. 211-240, Interscience Publishers, John Wiley, New York (1963).

84. M.J. Luton, R.A. Petkovic and J.J. Jonas - Kinetics of recovery and recrystallization in polycrystalline Cu, Acta Met. 28 (1980), 729-743.

SUBJECT INDEX

AUTHOR INDEX